Francis Bacon's Skeptical Recipes for New Knowledge

Jagdish Hattiangadi

Francis Bacon's Skeptical Recipes for New Knowledge

palgrave
macmillan

Jagdish Hattiangadi
Department of Philosophy
York University
Toronto, ON, Canada

ISBN 978-3-031-52584-1 ISBN 978-3-031-52585-8 (eBook)
https://doi.org/10.1007/978-3-031-52585-8

This Palgrave Macmillan imprint is published by the registered company Springer Nature Switzerland AG.
The registered company address is: Gewerbestrasse 11, 6330 Cham, Switzerland

Paper in this product is recyclable.

Preface

The preface is a personal account of how I became a keen reader of Francis Bacon's theory of induction, after a professional career denying that there is or can be any such thing. The idea of induction he propounded was not like Aristotle's *epagoge*. Nor was it like the inductive logic and probability studied since the eighteenth century, which has carried over to current literature. The different kind of induction he proposed was an application of *skeptical* ideas. It was my key to unlocking the outline of his proposal. I gleaned a new understanding of how natural philosophy, which adopted his method, gives us knowledge as the science of the world around us—which is knowledge and not just refuted hypotheses or conventions about reality. A summary of the book's central thesis will be found in this preface.

My understanding took a surprising turn late in my academic career. I came to believe that my most trusted assumption over my professional life was mistaken. Throughout my teaching and research, I had assumed that a method of induction to make scientific discoveries was impossible. No workable "how-to" instructions could produce the next scientific theory. Induction had seemed to me to be a futile wish of the past.

"Induction" in the philosophy of science refers to how we establish some general and abstract principle as true—that is, how we know a piece of scientific wisdom or a universal theory. It is often thought about as an inference using a kind of logic. Aristotle introduced us to *deductive* reasoning. A general principle, among the premises of a valid *deductive* argument, necessitates the truth of its conclusion. For instance, the general principle that all humans are mortal and the particular that Socrates was a human allows us to conclude that Socrates was mortal. The conclusion follows necessarily. In any such deduction, the general principle among the premises sheds light on the fact in the conclusion—we may say that it explains the fact. Isaac Newton explained many facts about the solar system by deducing them from the laws of motion and the principle of universal gravitation. The deductions may be long and challenging. But his conclusions follow validly from his premises.

But how did he know the law of universal gravitation as a premise? Some philosophers have supposed there must be an *inductive* method. It would be an inversion of the deductive method of logic—a way of deriving general principles from particular facts. But no general principles would follow of necessity from a few particular cases that we have observed. It seemed to me to be more straightforward to suppose, as have many philosophers in the twentieth century, that all general principles are guesses, hypotheses, or conjectures—in other words, that there is no general method to getting to know the truth of general principles.

Quite the opposite now appears to be true. The method of induction proposed by Francis Bacon is not what I had imagined it to be. It is not like Aristotle's version of it. Aristotle proposed a method of discovering first principles that were wedded to an ancient and now discarded metaphysical scheme of Forms, Ideas, and Essences. No one uses such a scheme as an assumption, at least, not in a natural science—not, for instance, in physics, chemistry, or biology. Nor is Baconian induction like an inductive *logic*, as modern philosophers have proposed. It is a way of knowing, but not by inference from some premises.

I came to appreciate Francis Bacon's real suggestion for the first time. He did have a new way of learning about the world. I had not seen it earlier because I had taught myself not to recognize it. With a slightly new reading of his remarks, I began to make sense of some improvements in modern science that I previously found hard to comprehend. The great change following a small improvement in understanding his text was an uncertain experience for me. I know that it is easy to read things into the texts of others and find there what you want to see. However, I need to tell my story. I now believe the story I will recount is true, at least in its general outline. But now I must await your critical response.

Francis Bacon proclaimed a new method for obtaining abstract knowledge in natural philosophy. He proposed it as an alternative to Aristotle's method of science, which consisted of observing things under normal conditions and then extracting knowledge of their essence from the record of those observations. Francis Bacon advocated the use of experiments and recommended a novel manner of recording them. Before his time, experiments were rare. They were used sparingly by alchemists and a few mathematicians to peek into the true but esoteric natures of things. Experiments in alchemy played the same role as observations in Aristotelian natural philosophy, but they were conducted to support a different theory of the nature of things.

Francis Bacon found a new use for observations and experiments, which has made them central to modern science. His idea was to use them in what may be called a Socratic cross-examination—an *elenchus*—of nature. Elaborating on Francis Bacon's *elenchus* is a large part of my task. He made his new proposals in the early part of the seventeenth century before the beginning or the birth of modern science.[1] There is a lively discussion in the literature about what he contributed to modern science, because not everyone has understood what he advocated for experiments.

In a nutshell, his new idea was to use difficulties of facts indicating the nature of things to generate general theory, which is why he called it, following Aristotle, a kind of *induction*. It was an unfortunate choice of words. It is often confused with the earlier method. Francis Bacon's new idea was to use experiments to oppose observations, and to undermine them and each other. You generate discrepancies in the evidence. His induction is unusual: at first, you make things more difficult by problematizing the evidence of a nature. You record the problematized evidence in a new kind of "natural history." Then, just as its many conflicts appear almost impossible to comprehend, you solve the problem. He called this final step the "interpretation of nature," meaning something like decoding a cipher.

In this book, I call this step "solving a puzzle," departing from contemporary uses of the word. I treat a puzzle as composed of many individual problems. Each problem is important. Since the evidence itself is problematic, we cannot just throw away evidence refuted by a discrepancy in it. The old evidence and the new discrepancy are both recorded in an experimental natural history. Evidence is not like an opinion or a hypothesis. Therefore, he contrasted his new method of interpreting nature with the old "anticipations of the mind," hypotheses, opinions, or conjectures. When refuted, anticipations or hypotheses may be discarded without residue. You seek a new conjecture to replace the old. Anticipations of the mind may be very impressive when expounded by philosophers, especially if they can evade refutations by using various dodgy maneuvers. But when carefully collected evidence has discrepancies, we cannot just discard them. We need to understand why the discrepancies occur. We need to interpret the evidence in a new way. If there are many such discrepancies in a natural history, then it exhibits a difficult puzzle whose solution is an "Interpretation of Nature."

Francis Bacon wrote well. We continue to revere him as a great stylist of English literature (Vickers 1968). He expounded his new method eloquently and fervently. He was also well-read, and his prose erudite. However, he had no followers during his lifetime.[2] No one else anticipated that his method would be useful one day, though many seem to have found it intriguing. Not even his politically influential extended family members thought he was on to something. He became, in time, a very prominent member of the Royal Court in London and a Parliamentarian in the House of Lords. He was to become a famous statesman, rising to the rank of Lord Chancellor. Because he was well known and well read and wrote with style and passion, many learned people in Europe read his philosophy and reflected on his method. And yet, his academic influence was limited. For all his fame and power, the method he tried to promote all his life was not adopted by any philosophers—at least not openly while he was alive.

Then, more than thirty years after he died, a society was formed of fifty-five philosophers, which became the Royal Society of London. Their community acknowledged that they would use a method based on Francis Bacon's philosophy. The method was used with minor improvements to study diverse subjects, such as chemistry, optics, physiology, atmospheric physics, and astronomy. They called this new method their "experimental philosophy." We must use experiments to problematize the observations (and other experiments) to produce almost insoluble puzzles before finding unique solutions. The members of the Society made notable contributions in the first four fields using their slightly modified form of Baconian induction (Sprat 1667).

In those days, however, the grand prize for philosophy was in physics and astronomy, that is to say, in understanding the motion of planets. There was a need for clarity in the fundamental theory of motion and force, where many paradoxes seemed to abound. Basic questions had been left unanswered beyond what seemed to some even knowable at the time. It was a Baconian natural history that had sprung up without a plan to do so. Did the earth move, rotating on its axis and going around the sun, as the Copernican hypothesis has it? Or does the sun go around the stationary earth, as the ancient and venerable system of planets had it, which was still the common opinion? How were comets, tides, and the motion of things on earth to be understood? Everything connected to this Copernican dispute seemed murky. It was even harder to grasp all the puzzling items as a whole system.

Moreover, an Inquisition in Rome placed the most famous philosopher proclaiming the Copernican Hypothesis under house arrest. We must remember that the Roman Catholic Church at the time was considered the more tolerant religious institution when compared to the Protestant Churches following Lutheran or Calvinist teachings. Galileo Galilei's trial was, therefore, a shocking development. Finding true answers to his questions was undoubtedly among the most consequential of investigations.

Isaac Newton was able to work it out. He seemed to bring order out of chaos. He claimed to extract a new model of the planetary system from the facts by applying the methods of "experimental philosophy". He discovered universal gravitation, which principle was widely proclaimed as revealing a new force of nature. Moreover, Isaac Newton claimed that he used *experimental philosophy* in his astronomical discoveries to show that the Copernican system of moving earth is more nearly correct. The older model of stationary earth at the center of the

Universe was no longer tenable in astronomy. He was able to show how the planets moved in elliptical orbits, more or less, somewhat as Johannes Kepler had mapped them out. He also had a superior theory of the tides to the theories of earlier natural philosophers. His laws of motion could also subsume Galileo's law of projectile motion as part of this mathematical model. Odd as that may sound to us, he claimed to find out all this and more by "deducing" the theory of gravity from the phenomena. Even his claim to apply experimental philosophy sounds very odd to us since he could not play around with or experiment with planets or stars. What could he have meant by his methodological claims?

Newtonian physics became famous. After about a century, it became the science to emulate by other philosophers and in other sciences. The new physics was extolled for its virtue even by philosophers who sympathized with earlier adversaries of his experimental method. Some enthusiasts were soon celebrating the new approach to knowledge as a panacea: some optimistic philosophers even hoped to extend its application to find all fundamental truths in all subjects and solve all social and political problems. All this was part of an "enlightenment," inspired and made possible by Isaac Newton's famous method.

There were critics of the Newtonian method as it came to be identified. The critics said the method is invalid—impossible to implement because we cannot use it. Their arguments against induction were logical and clear: You cannot validly infer a general claim covering all places and all times from a few particular and local facts. Many philosophers, a minority of all philosophers at first, perhaps, were convinced that such a method could not work. Induction is a sham, they concluded. Enlightenment is just a wishful fantasy. Many philosophers were surprised by the arguments against the new philosophy. They were convinced by what happens in their science that it works. They continued to use it.

David Hume was the most notable early critic. I must respond to his clear and straightforward critical arguments as an important part of my account of induction. I will have to show where he erred, though I must ask you to wait until much later for my full diagnosis. However, I will not interpret Newton's physical astronomy as a Baconian induction in this volume, being too large a task, but leave his claim for elaboration in a subsequent treatment.

Philosophy came to be split in two. In one group were the opticians, astronomers, chemists, physiologists, etc. In small groups, they employed Baconian induction, or the experimental philosophy, gathered together as professionals. They continued to produce results in their respective subjects to much acclaim. They neglected the doubts and the arguments of the critics of induction. Later, in the mid-nineteenth century, these dogged practitioners of induction—practicing it in the face of Humean doubt—were given the collective title of "scientists." The other group of philosophers, who doubted that there was a valid experimental method of the kind that Francis Bacon, Robert Boyle, and Isaac Newton had claimed, continued as a group called "philosophers." For the most part, philosophers admired what scientists did, and some even contributed to science. Most scientists found philosophers, especially their theory of knowledge, amusing. The philosophers denied that there is a method, a recipe, a cookie cutter, a formula for doing science, which the scientists seemed to employ all the same.

It was astonishing to some philosophers that those other philosophers, who were later called "scientists," could practice their inductive or experimental method so successfully. They did get excellent results, true enough. But no one could state the inferential technique supposedly being used, at least not well enough to meet the known objections, such as those raised by David Hume. "How to do science" became a natural skill, a kind of tacit procedural knowledge. The gift of the scientist traveled from one lab coat to another. Apprenticeship became the mode of transfer of the new method. Induction—as a theory, if there was ever a coherent theory stated—went underground.

The methodology of science, or attempts to state how science succeeds, became enigmatic. Less than a hundred years after Isaac Newton's death, philosophers agreed that science did not advance using experimental philosophy, though scientists claimed to use it. The general claim was that the new method was empiricism, and David Hume had shown its limitations. The denial of a true method was soon to become widespread among philosophers. Despite that, they applauded Isaac Newton's great achievement in physical astronomy. They took up the task of enunciating the proper hidden method of modern science. William Whewell, John Stuart Mill, Stanley Jevons, and Charles Peirce were famous thinkers who thought deeply about these matters. If there is a method, their effort to divine it became a distinct and unrelated enterprise—unrelated, that is, from the task of the particular sciences, because it seemed to fall to the critics of induction to show how science does what it does. If scientists do not get their admirable results by using the "experimental philosophy" of Isaac Newton, then how does it succeed? A few historians of the philosophy of science remember these seminal thinkers of the early nineteenth century. Their opinion became the backdrop of later thoughts on science.

And yet, if the method of science has always included experimental philosophy, or a skeptical evaluation of the evidence, as I have come to accept, then it has not necessarily and always been tacit. It had once been proposed explicitly by Francis Bacon. Robert Boyle had published copious notes commenting on aspects of the method. Isaac Newton had also written some very brief comments on it. In the eighteenth century, many optimistic philosophers had hailed it as a cure for all human social ills, before later philosophers abandoned it as quite absurd.

In the nineteenth century, two developments have since then commanded our attention. Each of them makes skeptical science or experimental philosophy harder to recognize. The knowledge gained in natural philosophy, in the form of models of reality, multiplied in number. That was the first development. These interrelated new models also dealt with everyday things. Ordinary things seemed to be better manipulated by using these models. Each model also became better calibrated and integrated, with increasingly better measurements of whatever could be better measured. *Natural science* (or the sum of these useful models) was distinguished as a different subject from *philosophy*. An account of this story is reserved for the second volume.

A second development is also evident in the nineteenth century. Entirely new sciences arose, namely the psychological and social sciences, which relied only on the ancient method of hypothesis, without a skeptical examination of the evidence. They were new in one respect. Each new science has an "empirical basis." The empirical basis cannot be purely empirical because sensory inputs are too remote from any of the objects under study. But the mock-empirical or quasi-empirical basis is a defining feature of each new school of thought or style of science. Each accepts a set of techniques and a way of formulating the quasi-empirical basis, which satisfies its exponents that the quasi-empirical basis is free of error or bias. For this reason, they may be called "probative" sciences instead of the skeptical sciences, which is my task to describe. This development, too, is reserved for the second volume.

When these new sciences arose, they arose as an application of what many thought was the method of the natural sciences, of physics, chemistry, and geology, for instance. But it is my thesis that they are different, though I will not address the new empirical sciences in any detail. They are well known to you. The historical phenomenon of the increasing density of knowledge in later science and the relation between the skeptical sciences and the probative sciences will only be mentioned briefly at the very end of my writing in the next volume, when I consider why the method of skeptical knowledge has been forgotten. In short, it is because the two methods have been confused as one.

William Whewell, who invented the word "scientists" to apply to philosophers specializing in the various sciences, also coined the phrase "philosophers of science" to describe the philosophers perplexed by the continuing success of an impossible method. He addressed the question of how scientists succeed in the art of discovery. He wrote in the middle of the nineteenth century when the density of knowledge in natural philosophy was getting widely recognized. We may wish to note that the distinction between two kinds of study, namely, a philosophical kind and a scientific kind, is presupposed in the expression "philosophy of science." My writing sprang from this field of study.

Oddly, my writing must also fall beyond its scope. If I am right, then there is no such thing as the philosophy of science because the two subjects are part of or blend into one another. William Whewell supposed that there is a certain kind of comparatively unreflective application of a method ("science") and another kind of reflective oversight of it (its "philosophy") that are each fundamentally different from the other. The same person may well do both, as did John Herschel and William Whewell, who were good friends. But it became a new article of faith, captured in the very expression "philosophy of science," that there is an essential difference between philosophy and science, even if one person can do them both. If the two are not so different, except in matters of detail, then "philosophy of science" would become "science of science" or "philosophy of philosophy." Nothing like that was in William Whewell's mind.

The boundary between science and philosophy is historical—a half-truth that has become ossified in our language. There is no real difference, but the perceived difference arose as follows. Once, there was only one subject called philosophy devoted to studying abstract knowledge of what was sought as wisdom. In the Middle Ages in Europe and west Asia, this changed. Among the Abrahamic religions, ethics and the wisdom associated with the goals of one's life (e.g., salvation) were reserved entirely for biblical theology. Natural philosophy excludes everything jealously guarded by theologians in the name of a jealous God.

A small part of an old subject (philosophy) was eventually freed for old-fashioned philosophical reflection. In natural philosophy, it is here that the renowned advances of modern science were made. Advanced scientific models have discovered hidden realities revealed in ordinary things as they appear to us. The discoveries also give us new powers to predict and manipulate them. In the seventeenth century, such models were considered a part of natural philosophy. By the nineteenth century, so many such models in different sciences had been discovered that we acquired a great density of new and practical knowledge about everyday things. These interconnected models were identified as "modern science" and contrasted with the rest of philosophy. Philosophy had never previously given us knowledge of that kind. Science soon became identified with the activity of the scientists—whatever they do to study the phenomena and advance their prediction by using their formidable models.

If the method of science is not about how to extract general knowledge from infallible accounts of experience, as the critics of induction claimed, then what is it, one may ask? At the beginning of the twentieth century, the work of Henri Poincaré and Pierre Duhem made it clear: the true and only method of science is the "method of hypothesis," as Plato had called it, or the "hypothetico-deductive" method as we now call it. The method is this. You propose an abstract hypothesis. A hypothesis about a thing targets a truth about it. But a hypothesis is like that shot in the dark, where you do not know what truth the shot has hit. There are many possible hypotheses, of which you have chosen this one. You could shoot in any direction or choose any hypothesis. Only one of these alternatives may describe the full reality of even a little part of the universe we are addressing. But which hypothesis could it be? Not knowing this, our guess is not known to be true even if we have confidence that it explains all the known facts. There are always many facts we have yet to be ascertained. As we discover them, we will modify which hypothesis is acceptable, as we advance into an endless future.

Only a very fortunate shot would actually hit the object we seek. Even if we do, we will not know that we have. Suppose we describe Isaac Newton's gravitational model of the solar system as a hypothesis. In that case, we are being very uncomplimentary to his model, which almost all physicists at the end of the nineteenth century would have sworn had hit its mark. Henri Poincaré was a brilliant mathematical physicist, too, and revered for his brilliant analysis of the Newtonian three-body problem. He claimed that when physicists say that the Newtonian principles of motion and gravitation are true, we must understand that they are only true by convention—a convention among mathematical physicists like him. We do not know anything of the kind. He thought this because we could assume a different principle and a different mathematical system to derive from them all the same facts of the Newtonian model (Poincaré 1905).

Pierre Duhem gave a brilliant critique of Isaac Newton's methodological claims. He argued that the latter could not have inductively established the physical laws governing planetary motion because he contradicted what was then known (Duhem 1954/1906). Pierre Duhem thus showed, in a logically convincing manner, that he could not have done this at all. His arguments went much beyond David Hume's original critique of induction, powerful though that critique already was. We could say that the principle of universal gravitation was serendipitous—just a lucky guess. But modern scientists seem serially fortunate. We seem to be making successful hits in many shots in the dark, too many for them to be a coincidence. Other academics in a University are not hitting as often with their darts. Science seems to have something special that makes its hypotheses lucky and its models useful. Someone should quickly bottle this something that makes them so blessed (I remind you, of course, that I think that Francis Bacon, Robert Boyle, and Isaac Newton had done just that, however imperfectly).

The apparent failure of a method of modern science, or induction, contrasts with the public reputation of science as very successful. If science is populated by hypothetical knowledge, how do we account for its remarkable accomplishments? We can divide the responses into three kinds. Some tried to reconstruct methodology in a new way. Some tried to account for the success by learning how to make do without induction. Some others tried to deny the accomplishments of modern science altogether so that they did not have to explain it. The three responses vie for our attention.

The first way led to an account of the inductive method which became *retrospective* in the nineteenth century. Methodology no longer *prescribed* so much as *recounted* how some had previously gained knowledge. Early in the twentieth century, methodologists agreed that this was our best. Many philosophers have defended induction in this retrospective form of it. For instance, many logical positivists in the twentieth century supported induction construed as a retrospective account. They invented systems of inductive inference, inductive logic, probabilistic analysis, etc. and used them brilliantly as analytical techniques for the rational reconstruction of some examples of past science. However, none of these analyses were employed in the science they offered as illustrations. As opposed to this construal of old science, we must note that Francis Bacon used the inductive method envisaged by *prescribed* scientists to follow in the future. When philosophers assumed that only retrospective analyses of methodology are possible, they assumed that a prospective inductive method has no hope. But Baconian induction is very much a prospectus for modern science. Modern science, according to one common reading of it, did not exist before the Baconian method, which helped it along the way. If so, then retrospective induction is not adequate to describe modern science.

Karl Popper, like Henri Poincaré and Pierre Duhem, was emphatic in his rejection of induction. His normative theory of science is about how to live with the disappointment that there is no scientific method, even of a retrospective kind. He accepted Pierre Duhem's trenchant critique of the inductive method and even extended it further (Popper 2002/1959). He also noted this weakness in the method of hypothesis, pointed out earlier by Pierre Duhem: If we refute a system of hypotheses, we can always defend a favorite hypothesis within the whole system of hypotheses. We can deflect blame because the premises are jointly refuted, and we must choose which premise or premises to change (Popper 2002/1963). I believe Francis Bacon knew this as well as Karl Popper and Pierre Duhem. The ease of defending hypotheses was one prominent reason he argued against "Anticipations of the Mind." He promoted the Interpreting Nature as a method, which he called his new organon, because otherwise it is so easy to evade refutations. If we conclude with Pierre Duhem and Karl Popper that Francis Bacon's method of Interpreting Nature is a sham, then we must learn to live with the mere anticipations of nature in science.

Karl Popper wrote that we should accept any refutation of our main theory for what it is, namely, as a refutation, and not try to deflect the criticism. Tests in science typically check for consequences derived validly from a few claims. The main hypothesis that interests us is just one of them. If a few claims fail a test, it can be attributed to any of the argument's premises. Therefore, it is always possible to blame a failed test on some other assumption than the main hypothesis we are testing. Karl Popper claimed that we should not be interested in protecting some favorite theory in science. We are engaged in trying to find out what is the case. In the pursuit of truth, intellectual honesty demands that we do not shelter a hypothesis we like by finding another suspect.

If there is no Baconian method, then Karl Popper's rousing call for the pursuit of truth and intellectual honesty in science is perhaps the best we can do. But is it enough? What if deflecting criticism to defend the grand hypothesis is the right step? For all we know, it may be faithful to reality—and we cannot access reality in any other way than by way of hypothesis and its criticism. The facts can always be questioned, but so can the theories of science. Every possibility remains open. In social science, they call this a lack of closure. We have to make the decision to close off the debate. If Duhem is right, methods internal to science cannot provide closure. He proposed that use "good sense." Many social scientists concluded that closure is provided socially, by conventional decisions or by consensus. Perhaps even extra-scientific social factors may intrude, like budgets or social conflict.

My story is about a detail that has been missed too often: we do have a bit more to help us than our intellectual honesty, though Karl Popper's insistence on it is still valuable. That detail is Baconian induction. In my story, however, I do not assume an essential difference between philosophy and science. We can use the skeptical inductive technique more widely than we have. Moreover, so-called scientists may also be excellent philosophers. They are perfectly capable of making contributions to general questions about knowledge, cosmology, and ethics. Twentieth- and twenty-first-century assumptions of an essential difference between science and philosophy, which philosophers have made concerning science, constitute a mistaken opinion—a mistaken opinion, I contend, that became common among the overseers of science in the nineteenth century.

Those of my readers who are familiar with the writing of Thomas Kuhn on this subject will recognize his pervasive assumption of this supposed difference between philosophy and science. He expanded on it by contrasting blind (but productive) practitioners of the specialized sciences and ineffective but farsighted historians and philosophers of science. This insight is the hypothetical basis of his view of scientists following "paradigms." He claimed that practitioners of science emulate admirable or paradigmatic achievements. Scientists emulate them—with blinkers on. Blinkers exclude philosophical reflection during long periods of "normal" science, which follows the acceptance of a paradigm. Blinkers help followers concentrate on routine tasks (Kuhn 1962, 1963). Thomas Kuhn could see farther than the blinkered normal scientists in what the scientists are themselves doing. He could do this through his farsighted profession. He was now a historian and a philosopher of science, though his doctoral degree had been in physics.

His ideas became influential in the public domain. Perhaps this is because everyone can recognize the caricature of the scientist and of the historian/philosopher in his writing. By the time Thomas Kuhn wrote about it, it had been well over a hundred years that these stereotypes had come to occupy the public imagination: the scientist's emotionless rationalism and attention to tedious detail, as contrasted with the farsighted genius of enlightened philosophy.

If I am right, however, there really is a method of modern science. The method made natural philosophy, or science, successful. If so, then Thomas Kuhn captures a superficial fact about a small period in the history of science—about a period when scientists could not state how they were doing what they were doing so well. It was a time when philosophical knowledge became useful, and useful models acquired a new density. My story, once told, should help correct this deficiency. There is no need for scientists to be blind or philosophers to be ineffective. This Romantic contrast is entirely artificial.

Not all students of science concede that science has been successful. Taking a very long view, some might say that modern science is just another human intellectual endeavor. It supports a system of ideas at any time that is no better, relatively speaking, than many other systems of ideas in other cultures and at other times. The success of science is a rhetorical achievement, a bit of an illusion. We are the perpetrators of a fantasy and its victims, too. Being trapped in our times, we merely admire our current orthodoxy. We have talked ourselves into believing in our rhetorical flourishes. Science is not more or less successful than other adventures of ideas like it. Its success is whatever we decide to think about it to make us feel more important. This liberated conception of human intellectual effort was beautifully portrayed by the Romantic poets of the nineteenth century, by John Keats and Percy Shelley. Among philosophers of science it was revived more recently by Paul Feyerabend (1975).

We can insist that the success of science is itself actively constructed by us and not just discovered. We must admit that, if we have constructed that reality, we have built it very well. Few people outside of academia will point to Science and Technology Studies as a great and successful endeavor or Philosophy of Science as an admirably successful subject, for that matter. But everybody believes that science has given us something practical and substantial, whether they love it or hate it. Let me put it as follows. We, students of the sciences, seem not to have constructed our realities as well as have the sciences theirs. Scientists seem to know what they are describing, whereas we seem to be floundering in our general description of science. Most observers will readily point to modern science as the more successful of the two. Among those who oppose science and technology, some see it posing a grave danger to human values or even to our survival. But they will admit that science and technology has been successful on its terms. Its success creates the hazards they point out. Science can give us useful knowledge, even if it tempts us to use it badly.

All the talk of the *social construction of reality* arises from our apparent inability to show how to find closure in scientific arguments. Many social scientists seem to think that we get closure only socially, by processes external to the science (i.e., the philosophy) itself. I suggest that one story we must also consider is mine: *Baconian science provides closure.* It also does something else that neither the philosophers of science nor the social scientists can explain. In the pursuit of abstract knowledge, we seem to spawn new and valuable technologies. How does technoscience even happen? We must appreciate Francis Bacon's idea of skeptical cross-examinations of nature from the very form of which new powers over nature can be gained.

Paul Feyerabend once said that no single scientific method can be prescribed for all science at all times—other than "anything goes." He also said that the philosophy of science is a subject with a great past. This description is a naughty dig at other philosophers of science, implying that their thoughts have no future. However, philosophy of science is not paradoxical because there is no single true method of science, as Paul Feyerabend claimed. It has no future because there is a method that always recommends itself. It is a philosophical method, which I will attempt to state as found in the writings of Francis Bacon. I will also illustrate how Isaac Newton exploited its features as a craftsman. He used the Baconian method of contributing to a problematized natural history of light and then solving the complex puzzle with his theory of white light as a mixture of colors.

But we will begin by looking at the difficulties of induction as ordinarily it is conceived.

Toronto, ON, Canada Jagdish Hattiangadi

References

Duhem, Pierre. 1954/1906. *The Aim and Structure of Physical Theory*. Originally, *La Théorie Physique, son objet et sa structure.* Tr. Philip P. Wiener. Princeton, NJ: Princeton University Press.

Feyerabend, Paul. 1975. *Against Method.* London: Verso.

Harvey, William. (1628) 1909. *On the Motion of the Heart and Blood in Animals.* Tr. Robert Willis, Vol. 38 Part 3 in *The Harvard Classics,* ed. Charles Eliot. New York, N. Y.: P. F. Collier and Son.

Kuhn, Thomas. 1962. *The Structure of Scientific Revolutions.* Chicago: University of Chicago Press.

Kuhn, Thomas. 1963. The Function of Dogma in Scientific Research. In *Scientific Change: Historical Studies in the Intellectual, Social and Technical Conditions for Scientific Discovery and Technical Invention, from Antiquity to the Present,* ed. A.C. Crombie, 347–69. London: Heinemann.

Poincaré, Henri. 1905. *Science and Hypothesis.* Tr. W.J. Greenstreet. London: Walter Scott.

Popper, Karl. (1959) 2002a. *The Logic of Scientific Discovery.* London: Routledge.

Popper, Karl. (1963) 2002b. *Conjectures and Refutations: The Growth of Scientific Knowledge*. London: Routledge.

Sprat, Thomas. 1667. *The History of the Royal Society of London, for the Improving of Natural Knowledge.* London: T.R.

Vickers, B. 1968. *Francis Bacon and Renaissance Prose.* Cambridge: Cambridge University Press.

Acknowledgments

In 1983, Gregory Vlastos published a celebrated paper on Plato's *Gorgias*, whose analysis informs my claim that Francis Bacon's method solved the problem that Plato failed to solve adequately, and which we now value in its application to modern science. Richard Popkin's history of early modern skepticism is an important background of my effort to understand Francis Bacon's defense of his religion from skeptical scrutiny, which I claim is a source of secular modern science. Karl Popper's realistic philosophy of science exerted a great influence on this book. His fallibilism and his realism raised difficulties for me and set the stage for the main problems I have addressed. These three philosophical influences will be seen throughout this book and its sequel, to be published, on Isaac Newton's physical astronomy and the strange subsequent disappearance of Francis Bacon's method from accounts of modern science.

The occasion leading to this turn in my research was an intriguing remark by Brian Baigrie. In 1998, returning from a research trip conducted for the study of Isaac Newton's unpublished papers at Cambridge University, he reported seeing many bins of papers marked "Observations and Experiments." He wondered why they were regularly mentioned together. I was at the time reading Niels Bohr's scientific correspondence and wondering why Albert Einstein kept deferring to Niels Bohr's interpretation of quantum mechanics as *the best account of the experiments.* This had not made sense to me, as a philosopher, because different interpretations of the same equations with a given empirical content should be regarded as *empirically equivalent.* What was it about experiments, I wondered, that is missing when restated as empirical observations? Brain got me started on a renewed study of Isaac Newton, Robert Boyle, Francis Bacon, and Plato's Socrates, as I went backward in time. This book traces my journey coming forward again, as I reinterpreted their inquiries skeptically.

A number of students and colleagues who read early versions of my opinions helped me improve them with their comments and criticism and also offered encouragement when encouragement was scarce. Slobodan Perovic, who was studying the development of quantum mechanics, helped me see many nuances in the application of Baconian induction to it. Others include Itamar Tzadok, Michael Stuart, Abigail Klassen, Melissa Elmaleh, John Fitzmaurice, Deivide Garcia Y Oliveira, Tyler Delmore, Kei Yoshida, Amanjit Babra, and John Buchanan, the last of whom also helped with the preparation of the first draft of the References and an Index supported by funds from a York University Small Research Grant. I thank Springer International for their generous permission to reprint in Chap. 9 much of the writing from *Boston Studies in the Philosophy and History of Science* v.335, ch.9. I also thank Hackett Publishing Company Inc. for their generous permission to reproduce René Descartes' optical diagrams in Chap. 11. The errors that remain after all the kind help are my responsibility: I assert copyright as the sole author of this book. But this assertion creates a well-known paradox of fallibilism, because I may be affirming and denying some statement at once without knowing which. Fallibilist paradoxes are not addressed adequately in this book or its sequel. I am grateful to Anandi Hattiangadi for prodding me to address them fully thereafter.

Notes

1. Throughout this volume, references to Francis Bacon's (1620) *New Organon* are made by citing the book number and the number of the aphorism in that book, as in "Bacon, II.17" for Book II aphorism 17. All quotes from it are taken from the Ellis translation in volume 4 of *The Works of Francis Bacon*, eds. James Spedding, R.L. Ellis, and D.D. Heath, trans. R.L. Ellis (Longman & Co., 1857–74). Other works are Bacon are cited normally, where possible from this same Spedding-Ellis-Heath edition.
2. A possible exception may have been William Harvey, though the evidence for the influence of Francis Bacon's thought on his theory of the circulation of the blood (Harvey 1909–14/1628) is weak.

About This Book

The book is about Francis Bacon's skeptical method and successful application to natural philosophy in the seventeenth century. This volume contributes to the history and philosophy of modern science, which, together with its planned sequel on Isaac Newton's physical astronomy and John Locke's empiricism, tells the story of the emergence of modern science as we know it out of natural philosophy. Though the seventeenth century is its primary focus, it traces the roots of Francis Bacon's method of science to ancient skepticism, as practiced by Socrates, Plato, the Academics, and the Pyrrhonians. Historians of ancient philosophy and science would be interested in examining its claims. Francis Bacon recommended a new Organon against the background of Anglican theology, which, more than the Puritan or the Roman Catholic, shaped his thought. Theologians will be interested in this aspect of the present and subsequent work. The claim that is made in this book is that science produces knowledge and not just speculations on which there is intersubjective agreement. My claims will interest philosophers generally, but more particularly epistemologists. The thesis will also engage the interest of those studying the sociology, anthropology, or ethnomethodology of science who also study such epistemological questions from their vantage point. Most importantly, perhaps, it should be of interest to scientists who take an interest in understanding the science to which they contribute and the general educated public following these remarkable advances of modern science for the last three hundred years.

Contents

1 **Logical Doubts Concerning Induction** 1

2 **Socrates, and the Skeptical Craft** 31

3 **The Active Fallibilism of a Situational Skeptic** 53

4 **Pyrrhonians, a School of Skeptics** 73

5 **Freedom by Confinement** 95

6 **The Sphinx** 117

7 **Evidence and Its Refutation** 133

8 **Francis Bacon's *Elenchus*** 155

9 **A Method for Mathematical Physics** 185

10 A Foundational Scepticism 213

11 Applying the Foundational Method 233

12 Applying The Experimental Philosophy 249

References 273

Index 281

List of Figures

Fig. 11.1	Reflection	236
Fig. 11.2	Refraction	237
Fig. 11.3	Total Internal Reflection	239
Fig. 11.4	Spectrum	242

1

Logical Doubts Concerning Induction

I

We may distinguish two schools of thought concerning the method of science. Both schools adopt hypotheses to model theorizing in science. One school also explicates induction as a scientific method of inference. If a deductive inference proceeds validly from a premise to a conclusion, then induction goes from many such conclusions backward to its premises, providing confirmation or inductive support. An inductive inference may be logical, or probabilistic, though not deductively valid. The second school argues against induction. They regard whatever the first school describes as either impossible or pointless. Their strategy is to make do without inductive argument.

The second school has hosted the more forceful critics of the modern scientific method. We must contend with the arguments of the second school if we are to defend the method of Isaac Newton (following and improving upon that of Robert Boyle and Francis Bacon before him.) The critics of induction oppose the very existence of what we wish to examine. We may hope that defenders of induction as a form of inference are our allies. It is not always so.

J. Hattiangadi, *Francis Bacon's Skeptical Recipes for New Knowledge*,
https://doi.org/10.1007/978-3-031-52585-8_1

Professional philosophers of science who argue for an inductive method in science do not generally support or argue for Isaac Newton's method or the one Francis Bacon recommended. No one in the profession of philosophy any longer recommends a set of techniques or recipes for doing science. Induction is defended instead as a theory of inference, or as a theory of probability, or of an account of subjective preference. Any such account of induction is retrospective. We tell ourselves how to justify, rationalize, or prefer already established scientific theory.

As opposed to that, Francis Bacon proposed an instrument (an *Organon*) that he thought would be useful for future scientists to apply in practice. When he enunciated it, there was no socially organized science of the sort he recommended. Therefore, his thoughts contained a guide for future scientists, much like a cookery book of recipes or how-to-do-it explanations. In other writings, he even gave mock demonstrations of what science would look like (Bacon 1626). There may have been more to the Baconian method than a recipe, but there was certainly not less.

He wrote the *New Organon* as a directive to train and employ natural philosophers (called "scientists" since the middle of the nineteenth century). He envisaged them as an arm of government. To pursue the description of reality, bit by bit, natural philosophers would slowly but systematically proceed from the base of the great pyramid of gainful knowledge. At this base, they record facts of natural history. They move carefully and gradually from the ground to the apex, where the ascent leads them to the true theoretical metaphysics. Then they would come down again, by reasoning cogently, back down through the sciences to the detailed description of phenomena, where they could also find practical applications of their hard-earned knowledge of nature. Francis Bacon called his new technique "true induction" instead of what he called a childish or puerile form of verbal learning from experience that Aristotle proposed. This new method was actually used, as some modern science practitioners claimed. But it was not practiced by a government department first, as Francis Bacon had envisaged. It was employed by a freely formed club of well-to-do philosophers, citing Francis Bacon as an inspiration. They formed the Royal Society of London. The Fellows referred to their new method as "experimental philosophy."

Isaac Newton credited his discovery of the force of universal gravitation to his application of "experimental philosophy," or the inductive method. He claimed to have "deduced" theoretical results about the solar system from the phenomena and then made them universal by "general induction" (Newton 1848/1726, 440–42 and 589). He also credited his success in optics to the experimental method (Newton 2012/1704, v. 2, 229). Most philosophers outside the Royal Society in those early days regarded applying this method as futile. The new approach came in for widespread admiration in later decades, mainly due to the success of Isaac Newton's science.

The following story of this method is extraordinary: it was widely hailed soon after Isaac Newton's death as a great discovery in its own right. Presumably, some philosophers practiced it successfully, as advertised. Yet, a hundred or so years after Newton's death, it was condemned by some specialists studying methods as an error of judgment on Isaac Newton's part. Within two hundred years, it came to be regarded, on logical grounds, as hopelessly naïve. Isaac Newton almost took Francis Bacon's recipes for granted, as it seemed to be an accurate account of how to do science. They were of practical use to him in his study. But we think such a method is not possible to implement successfully.

As opposed to this conception of a method or an Organon or an instrument, the "inductive method" among professionals in philosophy of science more recently is very different: it is a retrospective account of scientific theory. Philosophers have designed it to tell us how to interpret past scientific theories. They do not recommend or prescribe any particular methods for practicing science other than very general prescriptions (such as the rules of logic and possibly the importance of observations). Recent defenders of induction tell us instead how any particularly salient theory of science of the past can meet certain logical, probabilistic, or material criteria that they have now devised. Any such effort is analogous to a formal account in an analytic theory of valid inferences. A theory of logic tells us how to interpret a deductive argument as valid. So, too, does a theory of inductive inference. It shows us how to interpret a successful inductive argument as rational.

Consider an unusual recipe for making an omelet: "first, obtain the right tools and ingredients for the job; then proceed to make an omelet just as other cooks may do. If your omelet is well received, then we will provide a rational reconstruction of how you made it." This is the *rationally reconstructed omelet*. Of course, this recipe could extend to all the other items in a book of recipes, with very little change. All cooking would utilize identical recipes or techniques—with only small variations on those that we find for the rationally reconstructed omelet.

In the philosophy of science, both the defenders and the critics of induction in science in recent times agree that Francis Bacon's prospective method is inapplicable. Nor do they concede that offering another applicable way is their job. Deductive logic does not provide blueprints to construct geometric proofs, so why should inductive logic show how to practice science? The new recipe books only contain reconstructions of omelets from ages past, with no instructions for a new recipe, let us say, about how to coddle an egg.

Knowledge production and cooking are different, no doubt. Where recipes for culinary dishes are standard and expected, they are not for natural philosophy or science. Much depends on our conception of science and how it differs from other human products. However, the contrast between rational reconstructions and recipes is instructive. We see a limited value in any retrospective inductive account of science to understand a prospective recipe. We must study Francis Bacon's *New Organon* or its applications by Robert Boyle and Isaac Newton on their terms.

Philosophers of science often rest their case on a distinction between the context of justification and the context of discovery. The context of discovery is the proposal of a hypothesis. And after that, the context of justification is its test. If we follow this order, then we have, in effect, adopted the hypothetico-deductive method. In this method, when we propose a hypothesis or generalization, we must use our imagination and whatever assists us to come up with our guess. Hypotheses are fashioned or formed in the context of discovery. There is no method of discovery. We may then test, refute, confirm, or evaluate an invented idea. This effort falls within the context of justification. Philosophers of science can analyze it.

A discernible gap between the generation of a hypothesis and its evaluation must exist. Only then is there a distinction between the two contexts. Then, a hypothesis can be evaluated by neglecting how it was generated. Recipe books, by way of contrast, are for the generation of omelets. But we are concerned as philosophers only with its justification.

We will see that the study of prospective methodology is different. The tests are often found before the new hypothesis that they generate.

Philosophers of science have expended great effort in defending various forms of induction. They have given us multiple forms of inductive logic of ampliative inference. Most of these accounts are based on a model of the much-admired theory of a formal deductive logic proposed by Gottlob Frege (1953/1884) for the foundations of Arithmetic, and by Bertrand Russell and Alfred North Whitehead for working out in detail the logical foundations of all of mathematics (Russell and Whitehead 1910–13). We can understand inductive inferences as rational in many different ways. Most of them involve the use of probability calculus. It is often recast as a generalization of logical inference.

A general hypothesis (covering an infinite number of particulars) cannot follow as a valid *deductive* inference from the meager premises provided by the evidence, which consists of some particulars. But we can see that inferences are rational in different ways. For finite cases, we can modify sampling theory for induction. We can track changing subjective probabilities (on Bayesian accounts) for others to exhibit rationality. We will not be concerned about any of them. They are all the same to us in that they are retrospective accounts.

Suppose that one such account convinces us. Suppose we find that a certain received theory was accepted rationally. Its acceptance would show us something about the nature of reason, perhaps. But what would it show us about science? When the scientists accepted that theory, they did so without recourse to the philosopher's reconstruction. To that extent, science would be less than perfectly rational. The reconstruction is unlikely to change that fact, given that it is a retrospective account. We also note that our lack of an agreed-upon inductive logic has not deterred scientists, so far, from making advances in science. So, we will feel good that science is rational if we find a rational account of its choices. However, if it turns out that science actually exhibits an element of irrationality, then this would not change the outcome so far as science is concerned. It could only modify our estimate of the value of reason. Theories of inductive inference, therefore, need not delay us when we seek the method of science. It would be different if we were primarily investigating reason.

Francis Bacon proposed a method for pursuing knowledge that could not have been like a probability theory. It was a collection of recipes. Therefore, the many critics of induction are more pertinent to our study. Their logical negativism concerning method forms a familiar backdrop to the modern philosophy of science. We learn from them the many hurdles we must overcome to understand science in the seventeenth century.

II

Philosophers of science agree, for the most part, that David Hume, writing in the eighteenth century, found the main difficulty, or a form of one main difficulty, for any account of a rational method for gaining knowledge. David Hume formulated two closely related arguments that are influential. In one argument, he challenged the cogency of our reasons for knowing any future event. The other argument concerned the discernment of cause and effect and how possibly we could know them. David Hume's writing is so lucid that he makes a difficult argument easy to follow. Familiarity with his thoughts also makes later philosophers easier to follow. I will begin by quoting from David Hume.

"All the objects of human reason or enquiry may be naturally divided into two kinds, to wit, *Relations of Ideas* and *Matters of Fact*. Of the first kind are the sciences of Geometry, Algebra and Arithmetic; and in short, every affirmation, which is either intuitively or demonstratively certain," claims David Hume (Hume Part IV, paragraph 1). On his understanding, intuitively certain statements are those that are obviously true to someone who understands them at all, such as that two plus three is five, or that there is one and only one straight line that can be drawn between any two points. He calls these "intuitively certain" because certainty is a property of whatever follows with logical validity from intuitively certain premises, such as the theorems of Euclid's geometry from (presumably) self-evident axioms, or postulates.

He suggests that we can know mathematical truths by the mere operation of thought. "Matters of fact, which are the second object of human reason, are not explained in the same manner; nor is our evidence of their truth, however great, of a like nature with the foregoing. The contrary of every matter of fact is still possible, because it can never imply a contradiction…. *That the sun will not rise tomorrow* is no less intelligible a proposition, and implies no more a contradiction, than the affirmation *that it will rise*." (Hume 2011/1748, Part IV, paragraph 2).

If we are to record observations that the sun rises regularly, this is, of course, a record made in the past; and from the past if we were to reason by induction that it will rise tomorrow, we would need to assume that tomorrow (in the future) will be like yesterday (in the past). Reasoning alone will not prove that the future will be like the past (and we should say, "in this respect," which David Hume left out.)

We may suppose our knowledge of cause and effect, or our knowledge of how the solar system works, is enough to determine that the sun will rise tomorrow. David Hume concedes that we do reason in this manner, but he says that all our reasonings concerning matters of fact seem to be founded on the relation of *Cause and Effect*. Causal reasonings cannot be based on knowledge gained from the mere operation of thought. We need to seek the relation of a cause and its effect among matters of fact.

Here is the test to show this: Take an event or thing called a cause as given; we can imagine without self-contradiction that it has an effect different from the one that the cause is used to explain. The logical consistency of an alternate effect with the cause shows that mere thought will not tell us of the necessary relation between cause and effect. We can imagine that when this billiard ball hits that ball, the latter will not follow the laws of mechanics, but instead speed up the first, jump back, go straight up, or do any number of different things. When we imagine these things, we can do so without fear of contradicting ourselves. If we seem to know the relation of a cause to an effect, this also must be founded on experience. But this raises the further question, "*What is the foundation of all our conclusions from experience?*" (Hume 2011/1748, Part IV, paragraph 14).

"Suppose that we infer from experience certain hidden powers of a thing that assists us when we try to find out that it causes another. When a man says, *I have found, in all past instances, such sensible qualities conjoined with secret powers*: And when he says *similar sensible qualities will always be conjoined with similar secret powers*; he is not guilty of a tautology, nor are these propositions in any respect the same. You say that the one proposition is an inference from the other. But you must confess that the inference is not intuitive; nor is it demonstrative. Of what nature is it then? To say that it is experimental is begging the question. For all inferences from experience suppose as their foundation, that the future will resemble the past, and that similar powers will be conjoined with similar sensible qualities." (Hume 1748, Part IV, paragraph 21).

If David Hume has it right, then any inference that relies on a causal connection between separately experienced events (designated "cause" and "effect" respectively) also depends upon an inference that assumes that the future will be like the past. Since this principle cannot be itself known from the past, as we have already noted, we cannot cogently use causal reasoning to infer our knowledge from experience.

David Hume did not deny that we have knowledge of the world around us. He did not endorse that more stringent form of skepticism. His argument is about reasons: all the constant conjunctions of sensible qualities we have ever experienced are insufficient *as reasons* to inform us how sensible qualities will be conjoined in the future. Academic or Pyrrhonian skeptics may suggest, David Hume argued, that we do not know if the ground will support us, or that bread will nourish us. Since we need such beliefs to act in any manner, we would find ourselves paralyzed into inaction, entirely unable to act if we were to follow them, he thought. However, our need and ability to act would force us to give up these extreme forms of skepticism.

Springing to action, we would be obliged to act based on unreasoned assumptions. Like animals, we will be able to navigate our surroundings tolerably well. In this manner we will see that we do know some things based on experience that further rational enquiry cannot assure us; rather, they are known using the "reason of animals," that is, instinctively, or by custom or habit. He called his form of skepticism a "mitigated skepticism." He did not doubt that we know many things we rely on for managing our lives. He doubted that we can infer this great and practical knowledge from our sense experience by using our capacity for cogent reasoning. He challenged our reasons but not all of our knowledge (hence, his skepticism is mitigated.) Reason, he thought, was a slave of the passions: on this basis, he resurrected an account of instrumental reason (i.e., practical reason used for calculating personal advantage) that may be studied with interest.

David Hume's arguments are disquieting when we are studying methodology. All our knowledge concerning matters of fact must be gained instinctively rather than by reasoning. To animal instinct we may add human custom and personal habit, all of which is nowadays described summarily as *practices*, without making the sum any more rational. Nor does it mean we understand it. Our roots in human nature explain why we generalize as we do. Knowledge is generated mainly in the satisfaction of our wants, rather than from an inquiry into the nature of things. We cannot therefore improve upon instinct by a method of experimental reasoning. We must then condemn the new organon proposed by Francis Bacon, and practiced by Robert Boyle and Isaac Newton as impossible to follow unless we could re-interpret it as an extension of instinctual habits.

David Hume saw his work as experimental reasoning concerning human nature. His planned "science" of human nature, following, he says, Isaac Newton's lead, must therefore be recognized as a reflective account of what is manifest to a carefully skeptical mind. He does not infer any general psychological theories resulting from reasoning about our nature from facts gained through introspection.

To apply David Hume's argument to Isaac Newton's method of experimental philosophy is not farfetched. The difficulty with induction is this: any general statement entails that a large, possibly infinite, number of particular things that fall under it have a certain property. For instance, the consequences of Newton's third law of motion, "for every action, there is an equal and opposite reaction," are so diverse that we cannot observe more than a tiny fraction of such events. The law expresses how every possible interaction between any two bodies will compute. In our experience, however, we can have a record of only a few bodies close by recently, which may be perceived in a small and selected number of interactions.

Given limited experience, we suppose this law applies correctly to every interaction between any two bodies, anytime and anywhere. How can we reasonably claim to know that bodies far away, or far in the past or in the future, or bodies that are too small to observe, or too large to understand, or of a kind we have not yet studied, will yet obey this law? And if we do, how can we do so rationally *based on a study* of the few interacting bodies at our disposal? David Hume's critique extends naturally to all such cases. David Hume's *Treatise of Human Nature*, his first book on the subject, is unusual in how self-contained it is: Its object of study, namely human nature, is also, it is claimed, the basis, though not the rational basis, of our knowledge of it.

We could rationally ground the belief that the next instance of a generalization will be like past instances by relying on the principle that nature is uniform. But this claim is itself ungrounded. It is not a logical truism since its negation is not self-contradictory. Moreover, it cannot be grounded in experience by the method of induction or by abstraction, since this would make the reasoning involved flagrantly circular. David Hume suggests his "skeptical" solution, as he calls it (Hume 1748, Part V). He tells us that we know as much as we do because of a brute fact about us: we associate ideas in a way that makes us generalize as we do. This generalization is not a rational activity but a facet of human nature, much like animal instinct.

Isaac Newton's law of universal gravitation was not exempt from a skeptical interpretation of its basis. How could Isaac Newton have used reason when he discovered the law of gravity? We note, first, how the question in David Hume's hands has already become a matter of reconstructing past science. David Hume's analysed how Isaac Newton discovers this law: "There are some causes, which are entirely uniform and constant in producing a particular effect; and no instance has ever yet been found of any failure or irregularity in their operation. Fire has always burned, and water suffocated every human creature: The production of motion by impulse and gravity is an universal law, which has hitherto admitted of no exception." (Hume 2011/1748, Part VI, paragraph 4)

David Hume explains how we can know the laws of motion and gravity. Conjunctions of similar events occur regularly, without exception.

This particular regularity becomes deeply ingrained in us, by the law of association of ideas. This regularity, which we know by nature, is the law.

Let us call the natural or instinctive method by which we extract the concept of a causal relation from sensible phenomena the method of "natural abstraction." Let us call the deliberate and learned method of deriving general laws from particular instances the method of "inductive logic." David Hume's remarkable critique of rational inquiry may then be summed up in the following conclusion: natural abstraction, or the true experimental philosophy, while it is a real and natural process, is not a rational one, whereas inductive logic, or a rational or grounded experimental philosophy, is neither natural nor real.

III

Immanuel Kant was appalled by the skeptical analysis of causation on David Hume's analysis, even though he conceded its merit as a purely intellectual exercise. A moral agent, it would seem, should be able to cause her actions if she is free. Freedom to act as one decides seemed to Immanuel Kant to be central to the conception of a moral agent. Following the lead of Benedict Spinoza, he thought of a moral agent as an "autonomous" being, as someone who determines her laws, rather than obeying laws imposed from outside (which would make her "heteronomous," or less of a true moral agent than she could be.) This conception of moral autonomy made sense only if a person could, in truth, know that there are some things that she can cause to happen when she so decides. David Hume's analysis of causes and effects seemed inadequate to Immanuel Kant. Constant conjunctions are not causes. They could not be the basis of an adequate moral theory in which an agent must be the author of her actions. A world without known causes is amoral.

The skeptical argument used to support Hume's interpretation of cause is quite strong. But we think there *must* be a necessity in the causal relation. Such a necessity *must* go beyond observations of constant conjunctions. When a person gets up to go to work, the relationship between the decision to get up and the actual getting up is not a mere correlation that one observes about us. We seem to get up when we decide to go to work precisely *because* we chose to do so.

In his *Critique of Pure Reason* (1998/1787), Immanuel Kant tried to rescue the notion of causation from David Hume's skeptical critique. He conceded that the statement of the relation of a cause to its effect is not a logical law. It is not based on a mere relation of ideas. Its denial is not inconceivable, nor self-contradictory. We can imagine that when a billiard ball hits another, the other ball does not obey the law of impact. It may consistently behave peculiarly, so far as logic is concerned. The law of impact (Isaac Newton's third law) is not logically necessary. The question that comes up is whether there is some other necessary connection between them that is non-logical.

David Hume's skeptical analysis showed us that the relation of a cause to its effect is not logical and our knowledge of it must be based on our nature as human beings, i.e., that our knowledge of it has a subjective basis in human nature. But Immanuel Kant wondered whether something could come from us, without being a logical necessity, and still not be merely subjective. Could it be necessary in a different way, in which a cause may *necessarily* lead to its effect even though it would not follow from it by way of a *logical* necessity. A statement describing a cause to its effect may be necessary without its negation being self-contradictory. Such an anlysis would show how we, as moral agents, could be the authors of actions in a metaphysical sense.

Since the necessity of a causal relation cannot come from the mere observation of a conjunction of events called "cause" and "effect," the necessity must come from within us, somehow. Our nature must somehow impose this necessity on the causal relation, though this idea may seem strange. Let us call a logically true statement "analytic." A statement is "synthetic" if it is true by its agreement with some feature in the world. We seek a form of necessity that is not analytic but synthetic.[1] However, since the statement of the necessity of a causal relation was not derived from experience (i.e., it was not "*a posteriori*") it must be prior to experience ("*a priori.*") Immanuel Kant's quest, which is now well known, is to discover "how is a synthetic *a priori* judgment possible?"

Immanuel Kant began to study the peculiar necessity of mathematical statements in his attempt to answer this question. Perhaps David Hume dismissed them too easily as merely relations of ideas, or as "analytic" judgments. Once Immanuel Kant established to his satisfaction that the judgments of geometry and arithmetic are synthetic and *a priori*, he had saved ethics and politics, too. In that process, he had also arrived at a new understanding of how Newtonian science was related to the facts upon which it was said to be based.

Immanuel Kant suggested that geometrical necessity is not always analytic. He proposed that it describes the world of experience in a nontrivial way, even though it also possesses a certain peculiar kind of necessity. We need to follow this very subtle argument to see how David Hume's argument is situated in modern philosophy of science. We should note that this is not the kind of necessity that Aristotle found in the essences of things that give science its power to explain when invoked. Immanuel Kant suggests that it is found in Isaac Newton's use of geometry, arithmetic, and in causal arguments. This ingenious way of circumventing David Hume's famous critique is both subtle and interesting, even if it turns out to be flawed.

Our perceptions of the world can teach us about the world in two entirely different ways. In the more obvious way, we can learn from our experiences by paying attention to what is given in them, in the content of what we perceive. We may note that it is cloudy just now, though it was sunny this morning. This differential learning from perceptions is what David Hume understood by learning from experience. But there is another way we can learn from experience. By reflecting upon the totality of our perceptions, we may notice that every object of perception must exhibit certain features. We look at all our perceptions from above, so to speak, to get this. Immanuel Kant called such learning transcendental.

Arthur Eddington used a beguiling analogy that can be adopted to help us grasp this subtle point. Suppose that we go fishing in the ocean with a net. After catching fish for some time, and examining our catch regularly, we find that all the fish have a certain minimal size, let us say, a diameter of 2 inches. There could be a universal law of fish: none exist in the oceans that are less in width than 2 inches in diameter. But then we notice that our net, which we use to catch the fish, has a mesh whose holes are large; that we can only catch fish within this mesh that are larger than 2 inches in diameter. Now we know why our fish must (of necessity) have this minimum size. But this also shows us something else: we do not know if any uncaught fish are necessarily larger than 2 inches. Therefore, we may conclude that every *caught* fish obeys a necessary law. It obeys the law generated by our fishing net. But this is not a law of logical necessity, since we can conceive of smaller fish we do not catch that may or may not be there in the ocean (Eddington 1949, 16–17).

Immanuel Kant's idea is that any perceptual net has its "mesh," so to speak. We catch only objects in our experience that have, at a minimum, certain features determined by the perceptual net. Among the universal features of perceived objects is the feature that they are spatially extended. Their spatial extension, moreover, obeys the laws of Euclidean Geometry. All things in our perceptual net must always be ordered in time, thus obeying the laws of Arithmetical order. Perceived things must also obey the laws of causality. Geometry and Arithmetic are not logical necessities. The principle of causality is also not analytic. Physical objects, *when not caught in our perceptual net*, ("in themselves") could very possibly be otherwise. We can easily imagine a different kind of a world. This passes David Hume's test for it being not analytic. But it is necessary because we must perceive as we do.

To the extent that we can experience them at all, however, things *must* obey the laws of space, time and causality. In the last analysis, we cannot know physical objects except through perceptual experience. Our experiences must be of things that obey these laws. We no longer need to wonder why Isaac Newton's Geometrical analysis of conic sections could generate his model of the solar system. His Geometrical theorems are about the world (or synthetic) but known *a priori*—i.e., transcendentally, by reflecting on the spatial features of perception.

We can make a new use of this idea of mathematical and causal necessity. We can use it to connect those constant conjunctions in our sensations that David Hume found to be rationally inadmissible. It is not the relation of cause and effect alone that allows us to reason about hidden powers or forces, but mathematics applied in all its force that helps us find those secret forces, such as, for instance, the hidden force of gravity. Mathematics provides the means to reason about things to reach non-trivial conclusions, such as those of Newtonian physics. Mathematics is key to our reason, which transports us beyond animal instinct. Rational scientific inquiry is still possible, after all.

Or is it?

IV

That geometric truths are necessary ran into some difficulty a few decades after David Hume and Immanuel Kant had written their respective thoughts. A crisis was occasioned by the discovery of new geometrical systems that seemed to order spatial objects very differently than the old geometry of Euclid. Reflecting on them, Henri Poincaré developed a form of "conventionalism" as a philosophy of science. Conventionalism in different guises has had a profound influence on later thought. This development needs to be grasped to see why induction seems impossible to many philosophers of science.

Euclid had formulated some postulates or axioms from which all the geometrical truths would be derived. We can ask of his postulates whether they are like tautologies, i.e., analytically true. Are they self-evident or known merely by reflecting on the postulates themselves? If so, then we have made a good start. Further valid arguments can demonstrate the truth of all the theorems. If we can generate all the old geometry, then Euclid's geometry conforms to David Hume's remarks on how we know all its truths. The negation of any theorem would be self-contradictory.

One postulate of Euclid, however, is not so clearly an analytic truth, nor a tautology. It asserts there is only one parallel line to a given straight line. In more detail: given a straight line and a point lying outside it, there is one and only one straight line that can be drawn through the point such that it runs parallel to the given straight line. Two parallel lines cannot enclose a space. A mere operation of our thought does not show us this is true. Could it not be false? Just by thinking about it, how can we know that two straight lines that are locally parallel do not converge and cross at infinity? How do we know that they do not diverge from each other, allowing multiple parallel lines, at infinity? Infinity is a long way away. Moreover, mathematicians had learned a painful lesson: rules applicable to finite things may not apply to infinities. One can get paradoxical or self-contradictory consequences if we do. So, the parallel postulate may be true, but it is not a matter of intuitive certainty. How do we find out?

David Hume gave us a way out of our predicament. If our intuition fails us in the case of this postulate, we can use demonstration to assist us on the matter. If we assume the truth of the other postulates in Euclid's *Elements*, and the falsity of the parallel postulate, then, if David Hume is right, this should generate a contradiction. The truth of the postulates is a "mere relation of ideas." Geometry would be thoroughly tautological, just as David Hume suggests, if we found a contradiction when assuming the negation of the parallel postulate together with affirmations of the other postulates. If we do not find a contradiction, then perhaps Euclid's *Elements* does not have the analytic character that David Hume ascribed to mathematics.

Early in the nineteenth century, it was discovered that if several parallel lines can be drawn through a point outside a given straight line, then we still have a logically consistent geometry. It can be shown that if Euclid's geometry is consistent, then so is the new geometry. We can model all the truths of the strange geometry within the familiar one by a special translation of distinctive terms of the new System of Geometry to corresponding terms of the old. The theorems of this new geometry are very different, as they apply to things. But we can show that if the new and peculiar is inconsistent, then so is the established Euclidean geometry.

In Euclid's geometry, the sum of the angles of a triangle equal to the sum of two right angles. But in Nikolai Lobachevsky's geometry, the sum would be less, but different also because how much less than two right angles they are would depend on the triangle's size. The variation in size is unlike classical Euclidean geometry in which theorems about similar triangles can be proved without respect to their size.

We must take up Henri Poincare's pioneering reflections on "conventionalism." One does not know the truths of geometrical reasoning by a mere operation of thought. David Hume was mistaken on this point. Nor, as it turns out, are geometrical truths universally necessary, as Immanuel Kant had imagined. They do not have that force because we have different ways of reasoning spatially about our experience, each of which is cogent. It turns out that we have a *choice* in describing nature: we can assume either that there are parallel lines, or there are not, or many—all with equal internal consistency. Where Immanuel Kant proposed that our *synthetic a priori* knowledge of mathematics is a kind of necessity, Henri Poincaré (agreeing with the Kantian view in many ways) suggests that there is an inescapable choice to be made before any necessity can assert itself. Since knowledge and understanding require necessity, all our knowledge presupposes a choice made, even if we have made that choice inadvertently (for instance, by assuming the truth of Euclidean postulates.)

In fact, things are even worse, as Henri Poincaré noted. If we assume that no parallel lines can be drawn through a point outside a given straight line, or one only or merely two, or as many as you specify, up to an infinite number of parallel lines, then it can be shown that each one of these assumptions gives us a geometry that is internally consistent, if Euclidean geometry is itself consistent in the first place. We have a lot of choice.

"Ought we then to conclude that the truths of geometry are experimental truths? But we do not make experiments on ideal lines or ideal circles. We make them only on material objects. On what, therefore, would experiments serving as the foundations of our geometry be based? The answer is easy.... [we] constantly reason as if... geometrical figures behave like solids.... But a difficulty remains, and is insurmountable. If geometry were an experimental science, it would not be an exact science. It would be subject to continual revision.... *The geometrical axioms are therefore neither synthetic a priori intuitions nor experimental facts.* They are conventions. Our choice among all possible conventions is *guided* by experimental facts; but it remains *free* and is only limited by the necessity of avoiding every contradiction, and thus it is that the postulates may remain rigorously true even when the experimental facts which have determined their adoption are only approximate." (Poincaré 1905, 58–59).

In Henri Poincaré's judgment, when we abstract from experience, we do not get any knowledge that can rise to a level high enough to generate science. Science is not like a heap of facts, gathered from experience here and there. It has a reasoned structure. Abstraction does not give us a rational ordering of experience, which remains often like a heap until we subject it to the rigor of reasoning. In regarding abstraction as less than rational, he agrees with David Hume. Nor can we do physics without mathematical reasoning. In this, he agrees with Immanuel Kant. He defends his somewhat modified Kantian interpretation of reasoning in science, based on mathematics, and causality, without supposing that mathematics is universal and necessary. There are free choices that we must make. These choices take us beyond the highest level of a "heap" of abstracted knowledge. These choices form the very basis of reasoning in physics.

The adoption of a particular geometry, however, is not an isolated choice. When we choose a geometry appropriate to mechanics and physics, we are also obliged to choose appropriate laws of mechanics or causality that suit our geometry. To bring experiments and abstracted experience into harmony with it, we need to make changes to our laws of physics, too. Harmony is achieved by avoiding contradictions: We can choose theories freely—but we must avoid all self-contradiction. We need logically consistent *systems of hypotheses* in physics, which systems are the true units of choice, rather than pure geometries, or pure mechanics, or pure experiment laws, considered in isolation. These systems will have experimental and mathematical laws in them. That was one of Henri Poincaré's great contributions to physics and philosophy.

Henri Poincaré's methodological discovery is that physical theory has experimental consequences only when it is part of a whole system of hypotheses. A hypothesis is not at all useful to physics when considered in isolation. We may express this by saying that no experiments can condemn the choice of a geometry in isolation, though this way of stating the claim would be adopting certain expressions used by Pierre Duhem.

We may consider two different ways of doing science: one system with Euclidean Geometry and Newtonian mechanics, which is a familiar combination in classical physics; and another way with a modified (e.g., a four-dimensional) spacetime geometry, and a newly developed mechanics to suit this geometry, that would form a new system. Henri Poincaré was one of the most distinguished mathematicians of his day, and he could work out what would be needed by way of formal theory to make either choice work. Which form of physical reasoning would be best? That was his question.

In his book *Science and Hypothesis* (1905), he suggests that Euclidean Geometry is the simplest and Newtonian mechanics is the most convenient mechanical system of thought, though he makes each of these judgments in different parts of the book. He does not veer from his insistence that, eventually, the options are ours (provided we avoid contradictions.) How then do we choose? Convenience, or ease of use, is important to us. We also take the simplicity of the hypotheses into account. We may also prefer the beauty of a chosen system of thought. Such are factors that influence what we may choose. But criteria such as convenience, simplicity, or elegance can only influence us; they do not determine our choice. We must choose. That thought is at the heart of conventionalism.

Henri Poincaré's thoughts were quite influential, even outside of mathematics. They came to dominate concerns in the philosophy of science. In physics, though, they seem to have held him back. In his book, from which he has been quoted extensively, he considers every element of Albert Einstein's so-called "Special Theory of Relativity," later proposed in 1905. Henri Poincaré had already suggested in 1901 that we could postulate a four- dimensional space-time continuum, that we could consider Lorentz's transformation equations as a basis for a different mechanics, and had wondered about an alternative to Newtonian mechanics; he considered the difficulties of formulating an electromagnetic theory of light within Newtonian mechanics—in fact, anything that may be regarded as a new ingredient in Albert Einstein's special theory of relativity is also to be found considered in this book, and in other writings of Henri Poincaré prior to 1905 (e.g., in Poincaré 1901). Moreover, these matters are discussed open-mindedly, with great delicacy and concern for both the experimental facts known at the time and for the nuances of mathematical theory.

For all his mathematical brilliance and philosophical and physical erudition, Henri Poincaré did not assert the special theory of relativity itself, nor did he draw any consequences from it, such as the equivalence of mass and energy, as Albert Einstein did in 1905. It is hard not to conclude that his discovery of conventionalism, great as it was, held him back from grasping the special theory of relativity first.

Later, in *Science and Method* (1908), he changed his mind. We may prefer the old mechanics, and the old mathematics, he wrote, out of concern for the convenience and simplicity of each part. But if we consider the new mechanics as a whole, it is much simpler overall. But this choice of his was made after the fact. He had missed the theory of special relativity. He had opted instead for conventionalism.

V

Pierre Duhem generalized Henri Poincaré's methodological thought. In his now classic book, *The Aim and Structure of Physical Theory* (1954/1906), he claimed that the aim of physical theory is to summarize (unify) and logically classify (catalogue) experimental laws. Its aim is not to *explain* a group of experimental laws. That is to say, its aim is not to describe the reality behind appearances or to show their necessity. He sees the latter aim as equivalent to doing metaphysics, in which we try to describe a hidden (meta)physical reality. Science cannot be made subservient to metaphysics.

We read complex and counterintuitive theories in modern mathematical physics. They tempt us to ask what corresponds to the mathematical terms within them. Pierre Duhem thought that the attempt to interpret mathematical or theoretical physics as a description of reality (more recently called "scientific realism" among philosophers of science) leads to the adoption of a speculative metaphysical system. The physical thoughts that Galileo Galilei, René Descartes, or Isaac Newton propounded at different times were each a mixture of sensible physics and indefensible metaphysical speculation. He saw it as his task to describe the aim and structure of physical theory and leave out the metaphysical speculation from science (Duhem 1954/1906, 42–49).

Pierre Duhem wrote that speculative metaphysical systems are highly controversial, whereas there is considerable agreement over theory in physics. Therefore, by interpreting theory realistically, we would make agreeable physics subordinate to contentious metaphysics. Doing so would devalue the gradual and incremental gains made in physics with full agreement among physicists over the centuries. If we do not, we will mistake seventeenth-century physics for a contentious development.

There was, he thought, a change in the metaphysical speculation of some physicists that gave physics a bad name (i.e., of being revolutionary, as now we may say.). Physical theory unifies mathematically, and simplifies by its laws, many and diverse experimental facts, which may be derived from its laws. Unification happens gradually, and without controversy, as mathematics improves, and as experimental facts accumulate. We may speak in physics of hidden entities like instantaneous velocities, point masses, electric potentials, atoms, electrons, displacement currents, or forces and fields. Such seeming realities are described in terms found in physico-mathematical theory. Theoretical terms like these may may change radically from time to time, engendering much needless disagreement (Duhem 1954/1906, 9–18).

People disagree about such matters only when the mathematical equations of physics are too literally and naïvely interpreted. By trying to understand the world based on physical theory, we succumb to metaphysical speculations and think of science as undergoing dramatic change. Pierre Duhem's thesis is that when a collection of mathematical equations generates all the experimental results we need, it has fulfilled its aim—we have received all that we require from any physical theory. The controversy between the Church and Galileo Galilei's science is a case in point. It arises out of Galileo Galilei's confusion of speculative metaphysics with the proper aim of physical theory, which is to catalog and describe the known facts. We must not go beyond what is evident to what we do not and cannot know.

Pierre Duhem also noticed something remarkable in the interpretation of experiments. An experiment in physics is never a straightforward observation of phenomena. It is always also a theoretical (or mathematical) interpretation of it. Ordinarily, what we observe in the world is not mathematical at all. The result of an experiment in physics is an abstract and symbolic judgment. When we reason about physical experience, we use mathematics—as Immanuel Kant suggested and as Henri Poincaré concurred in his modified form of understanding physical inquiry. Pierre Duhem adds to their insights. A theoretical interpretation of the phenomena alone makes it possible to interpret them in terms of our abstract theory.

Even the use of instruments to measure quantities is possible only because we interpret the instruments symbolically in our theory. Instruments are part of a schematic model in physical theory. Only by this model can a physicist interpret the experiment meaningfully. The physical thing we manipulate in the laboratory differs from its symbolized model. Symbolic interpretation alone can give any experiment its relevance to physics (Duhem 1954/1906, 144–65).

Therefore, ordinary experience is only tangentially what science is about. It is really about the more precise results obtained from specialized experiments that it catalogues. When an experiment contradicts what a theory leads one to expect, it is not the observation, plain and simple, that confirms or contradicts the theory. The highly interpreted, symbolic experimental result confirms or refutes a physical theory. He draws some startling consequences for the structure of physical theory.

"[W]e must inquire very carefully into the theories which the physicist regards as established and which he used in interpreting the facts that he has observed. Without knowing these theories, it is impossible for us to understand the meaning that he gives to his own statements." (Duhem 1908, 159).

This may be called Pierre Duhem' s thesis of "the incommensurability of meaning between theoretical systems." He continues as follows, making his own meaning clearer, "If the theories admitted by this physicist are those we accept, and if we have agreed to follow the same rules in the interpretation of the same phenomena, we speak the same language and can understand each other. But that is not always the case. It is not so when we discuss the experiments of a physicist who does not belong to our school; and it is especially not so when we discuss the experiments of a physicist separated from us by fifty years, a century, or two centuries." (Duhem 1954/1906, 159).

"If on the contrary, we cannot obtain sufficient information about the theoretical ideas of the physicist whose experiments we are discussing, and if we fail to establish a correspondence between the symbols he has adopted and the symbols furnished by the theories we accept, the propositions through which that physicist translated the results of his experiments will be neither true nor false for us; they will be devoid of meaning…." (Duhem 1954/1906, 160).

Pierre Duhem's analysis of the structure of physical theory also demonstrates, to wide acclaim, that an experiment in physics can never condemn an isolated hypothesis, which is another celebrated result in the philosophy of science. An isolated hypothesis will have no experimental consequences if considered all by itself. To test a hypothesis, we must derive consequences from a whole system of hypotheses, of which we test a logical consequence. The experimental results need to be interpreted in the language of theory and in terms of a great many hypotheses of which it is a consequence. This generalizes a thought of Henri Poincaré's. One example that Pierre Duhem cites from physics will suffice.

"F. E. Neumann had suggested that in a ray of polarized light the vibration is parallel to the plane of polarization. O. Wiener showed how this was mistaken by deducing from Neumann's theory that if we cause a light beam that is reflected at 45 degrees from a plate of glass, to interfere with the incident beam polarized perpendicularly to the plane of incidence, there will appear dark and light interference bands parallel to the reflecting surface. He did the experiment and showed that they are not observed. Did he therefore refute the hypothesis in isolation?" (Duhem 1954/1906, 184). Here is his analysis:

> In order to predict the formation of bands in certain circumstances and to show that they did not appear, Wiener did not make use of merely the famous proposition of F. E. Neumann, the proposition which he wished to refute... but he used, besides this, propositions, laws and hypotheses constituting the optics commonly accepted: he admitted that light consists of simple periodic vibrations, that these vibrations are normal to the ray, that at each point the mean kinetic energy of the vibratory motion is a measure of intensity of light, that the more or less complete attack of the gelatin coating on a photographic plate indicates various degrees of this intensity. By joining these propositions, and many others that would take too long to enumerate, to Neumann's proposition, Wiener was able formulate a forecast and show that the experiment belied it.... [A]s H. Poincaré has shown, we can easily rescue Neumann's hypothesis from the grip of Wiener's experiment on condition that we abandon in exchange the hypothesis which takes the mean kinetic energy as the measure of light intensity.... (Duhem 1954/1906, 185–86)

To anyone not familiar with nineteenth century optical theory and experiment, no doubt, this detailed example will seem like an indecipherable Etruscan inscription. It demonstrates how incommensurability can afflict us. Pierre Duhem himself advised us how this could be so. Even so, his example allows us to imaginatively understand his sensible claim, that an experiment refutes no isolated hypothesis. In an attempted refutation of a hypothesis, we invoke a whole system of hypotheses to generate the proposition being tested. We also need to assume various interpretations to bring about the precisely needed experimental conditions to make the test possible. Given this fact in all its generality, however, we notice that, as a consequence, *we may always rescue the intended refutation of one hypothesis by substituting the abandonment of another in its stead*. Of course, to do so, we may need to make other changes in the system and thus avoid logical inconsistency. But we can always find ways to evade refutations. What a disaster this is.

Two theses are being claimed here; one thesis is that experimental derivations from theories are never made merely from isolated hypotheses, but always from a large number of hypotheses in conjunction, called the holistic conjunction thesis, or "holism." Willard Quine adapted and generalized what we know as Pierre Duhem's thesis, to cover any refutation of any theory even outside natural science. Physics does not suffer alone.

The second thesis is a logical truth: when a valid argument from some collection of premises has a false conclusion, then the collection of all those premises, considered conjointly, is false. Individually considered, any one of them or many of them, may be true. It is the nature of refutation to be holistic, logically speaking. Logic cannot determine for us which one of the premises is false when they are jointly refuted. We can call this logical phenomenon the "underdetermination" thesis. Refutations underdetermine the falsity of isolated premises in a refuted system.

Given holism and the underdetermination thesis, we can make two observations: one observation is that an experiment cannot condemn an isolated hypothesis; the second observation is that there is always a free choice to be made in choosing one's revised theoretical system. Our decision is made in the wake of an experimental refutation of theories, but it remains ours. We see that this is a general logical claim about any hypothesis—a generalization of Henri Poincaré's theory of free choice in theoretical physics. As Willard Quine put it, we can defend any hypothesis, come what may.

VI

Like Henri Poincaré, Pierre Duhem did not think that mathematical theory can be extracted from observed phenomena. But Pierre Duhem also argued systematically against Isaac Newton's claim to have made his discoveries by "deducing" them from the phenomena. Following Henri Poincaré, he argued against the possibility of an inductive theory of mathematical physics, as Isaac Newton claimed for his *System of the World* (v.3 of 1848/1726).

In the first place, physical theory goes well beyond anything that can be known to us from evidence. One may state this as an inference from past to the future, as David Hume did, or as an inference from a few cases to the many more, as Aristotle had formulated it. Whichever way we consider it, it is an "ampliative" inference as philosophers of science describe it today.

Pierre Duhem makes the distance between theory and observations much more distant than this. Physical theory is symbolic in nature. Consequently, its relation to any ordinary observations is very tenuous. Observations become relevant to physics only when they have been interpreted symbolically. Consequently, any theory of induction that suggests that we begin with observations and then "deduce" a theory cannot be correct, because by its very nature, symbolic accounts are different from our ordinary observations about things.

In physical theory, we are always going essentially beyond the observations, even while merely classifying experiments. We may find new kinds of phenomena that could not have even been suspected without the theory. Theories with such novel consequences are particularly highly prized. This would also seem to militate against theory venturing as little as possible beyond its factual or experimental inductive base. Novel consequences are novel precisely because they lie beyond any merely inductive inference of theory from the facts.

Pierre Duhem developed this argument in an interesting way. He pointed out that we value the content of physical theory especially when it leads to unexpected or novel consequences. Physical theory aims to give a unification and classification of the known experimental laws by using mathematical symbols. In doing so, we get a much more precise description of experimental regularities than we could obtain from ordinary observation. The symbols used in theory give the theoretical laws simplicity and beauty, as Henri Poincaré noted; but they also give the laws we write down a greater power of prediction than we could have extracted from ordinarily observed phenomena. Sometimes, the production of an abstract and symbolic theory will give rise to predictions of entirely new kinds of phenomena that stimulate new experimental investigations (Duhem 1954/1906, 293–301).

Augustin-Jean Fresnel, for instance, first presented his equations for a wave theory of light to account for diffraction in light, i.e., the phenomenon of blurred shadows—which were noted by Isaac Newton. But Augustin Fresnel's symbolic representation of light as a wave has a consequence that was not even known to the author. Denis Poisson showed that on one solution of his new wave equation, a beam of light passing by a circular object would project a spot of light right in the middle of its shadow on a screen behind the circular object. The screen must be placed at the right distance from the object and the source of the light (Fresnel 1900/1818). But who expected a spot of light in the middle of a shadow? The theory must be mistaken!

Poisson spots can be easily observed once we learn of them. They opened up a whole new experimental field for studying interference phenomena. But they show us clearly that Augustin Fresnel's theory could not have been "deduced" wholly from the experimental observations already made because they have these novel effects that he did not know and could not have known in advance. Suppose we value the new directions that physical theory indicates to us. In that case, we must recognize that it is not deduced validly from the previously known phenomena as Isaac Newton's "experimental philosophy" supposed.

A second argument is very closely related to this. The very exactness of the laws of physics makes them unsuitable for induction from any facts that can be observed. Unless we first interpret the facts in symbolic terms, the inference from facts to theory cannot be valid. Consider the law of gravity. It is exact, as formulated. The force of gravity between any two bodies is a certain constant multiplied by the product of their masses and divided by the square of the distance between them. None of these quantities is approximate, in theory. Not one of our ordinary observations, however carefully we go about them, can give us facts that match the exactness of theory. What we observe invariably has a lot of inexactitude. Therefore, our exact physical theory cannot be derived from the inexact and approximate observations of things we can record.

One conclusion that can be drawn from this fact is that mathematical laws of physics are the free creations of the human mind, as Albert Einstein described them at one time (Einstein and Infeld 1938, 33). They are our creation because they postulate exact relations between symbolic objects that are extracted from our experience, but which cannot be precisely enough known in experience; and this creation is free because, as Henri Poincaré showed, we have a choice regarding which symbolic theories we adopt, provided that we choose systems that avoid contradictions.

The third and last argument is perhaps the most devastating of all, and it is Pierre Duhem' s most telling contribution to the impossibility of a method of induction.

Pierre Duhem considers a method that would yield a proposition that is "either a law drawn from observation by the sole use of those two intellectual operations called induction and generalization, or else a corollary mathematically deduced from such laws.... It was this sort of physical theory that Newton had in mind when, in the General Scholium which crowns his *Principia*, he rejected so vigorously as outside of natural philosophy any hypothesis that induction did not extract from experiment; when he asserted that in a sound physics every proposition should be drawn from phenomena and generalized by induction." He continues, a little later, "Is not his theory of gravitation derived entirely from the laws which were revealed to Kepler by observation...?" (Duhem 1908, 190–91).

Pierre Duhem shows very carefully that it is indeed true that Isaac Newton relied on the first and third of the three laws of Johannes Kepler to derive certain consequences of value to the theory of universal gravitation. Nevertheless, he argues forcefully for the following extraordinary conclusion.

"The principle of universal gravity, very far from being derivable by generalization and induction from the observational laws of Kepler, formally contradict these laws. If Newton's theory is correct, Kepler's laws are necessarily false." (Duhem 1908, 193).

The argument for this claim seems to be quite unassailable. From the Newtonian system of physics, including the theory of gravity, we can derive consequences that violate the laws that Johannes Kepler had stated. Planets do not move exactly in ellipses, but only approximately so. Two planets will affect each others' paths by gravity, which throws them off their elliptical orbits. This deviation is more noticeable when planets are closer together. Isaac Newton had himself described how Saturn's orbit was peturbed when it was near Jupiter's. When Isaac Newton proposed his new account of the planets, Johannes Kepler's laws became interpretable as true at best to an approximation, in a model in which only one planet and the sun are being considered mathematically in isolation. Kepler's laws are strictly false—each of them, as shown by derivation from Newton's laws.

Karl Popper extended this argument in his influential paper, "The Aim of Science." Isaac Newton's new law also contradicts Galileo Galilei's account of the parabolic path of projectiles on earth (Popper 1983, 131–49). In his account, he claims that a "parabola is really a section of a very elongated ellipse, which is mathematically a different thing altogether." The insight that a parabola and an elongated ellipse are *approximately the same* does not change the fact that they are *mathematically incompatible* shapes. They look similar. Isaac Newton's account was far deeper and could explain why planets and satellites (and projectiles on earth) appear to move according to the very same previously stated laws. But it could also predict under which conditions we can observe that they do not obey those phenomenal laws. When they do not, a planet may make merely a slight deviation, but it is a deviation that we nevertheless value as a remarkable consequence of Isaac Newton's theory. It convinces us that Isaac Newton had something of inestimable value to tell us.

And yet, it follows from this very fact that Isaac Newton could not have derived his laws from those of Johannes Kepler or Galileo Galilei, or both. If his general theory is true, then the more particular or the so-called observational laws of the latter are false. How can we "deduce" a law from the phenomena (as Newton claimed) even as we are in the process of correcting, or contradicting the phenomenal laws from which we are "deducing" it? Devastating objections to the claim for induction have now been formulated. We must answer them if Francis Bacon and Isaac Newton's method is to be rescued.

What, then, are we to make of Isaac Newton's claim, that he was an "experimental philosopher?" Was he following Francis Bacon's new inductive method? Was he using the method of hypothesis, though he was castigating others for failing to appreciate his inductive method? Was there another method altogether—of "experimental philosophy"—which could meet the devastating criticisms we have reviewed? We need to see for ourselves.

Note

1. Immanuel Kant's as well as David Hume's concerns with the epistemology of mathematics spring from John Locke's form of empiricism to which they subscribed. The importance of empiricism for our amnesia concerning Baconian induction is not addressed in this book, but found in its sequel, to be published. Understanding the necessity of mathematical knowledge is a difficulty for John Locke, with which he does not adequately cope. Emily Carson (2006) has provided a detailed account and guide to show how Immanuel Kant's conception of mathematics as synthetic though *a priori* arose out of his attempt to solve a Lockean problem for empiricism.

References

Bacon, Francis. 1626. *Sylva Sylvarum, or A Naturall History in Ten Centuries.* Edited by William Rawley. London: William Lee.

Carson, Emily. 2006. Locke and Kant on Mathematical Knowledge. In *Intuition and the Axiomatic Method*, ed. Emily Carson and Renate Huber. Dordrecht: Springer.

Duhem, Pierre. (1906) 1954. *The Aim and Structure of Physical Theory.* Originally, *La Théorie Physique, son objet et sa structure.* Tr. Philip P. Wiener. Princeton, NJ: Princeton University Press.

Eddington, Arthur. 1949. *The Philosophy of Physical Science.* Cambridge: Cambridge University Press.

Einstein, Albert, and Leopold Infeld. 1938. *The Evolution of Physics*. London: Cambridge University Press.

Frege, Gottlob. (1884) 1953. *The Foundations of Arithmetic: A Logico-mathematical Enquiry into the Concept of Number*. Tr. J.L. Austin. Oxford: Blackwell.

Fresnel, Augustin. (1818) 1900. Memoir on the Diffraction of Light. In *The Wave Theory of Light: Memoirs by Huygens, Young and Fresnel*, ed. Henry Crew, 79–144. New York: American Book Company.

Hume, David. (1748) 2011. *An Enquiry Concerning Human Understanding* ed. Eric Steinberg, Hackett Classics, Indianapolis, Indiana.

Kant, Immanuel. (1787) 1998. *Critique of Pure Reason*. Tr. Paul Guyer, Allen Wood. Cambridge: Cambridge University Press.

Newton, Isaac. (1726) 1848. *Newton's Principia: The Mathematical Principles of Natural Philosophy*. Tr. Andrew Motte. New York, N.Y.: Daniel Adee.

———. 2012. *The Optical Papers of Isaac Newton*. 3 vols. Tr., ed., Alan E. Shapiro. Cambridge: Cambridge University Press.

Pappus, 2011/ 4th c. CE, v. 7 of Collection. Tr., ed., Alexander Jones, New York, N.Y. Springer.

Poincaré, Henri. 1901. *Électricité et optique. La lumière et les théories électrodynamiques; leçons professées à la Sorbonne en 1888, 1890 et 1899*. Paris: G. Carré et C. Naud.

———. 1905. *Science and Hypothesis*. Translated by W.J. Greenstreet. London: Walter Scott.

———. 1908. *Science and Method*. Translated by Francis Maitland. London: Thomas Nelson & Sons.

Popper, Karl. 1983. *Realism and the Aim of Science*. Edited by W.W. Bartley, III. London: Hutchinson.

Russell, Bertrand, and Whitehead, Alfred North. 1910–13. *Principia Mathematica*. 3 vols. Cambridge: Cambridge University Press.

2

Socrates, and the Skeptical Craft

I

Francis Bacon's method draws from three different kinds of skepticism. First and foremost, he draws from Socrates, who claimed he did not know what wisdom is, though he was constantly searching for it. In such a search and in his inductive method, he identifies with Socrates. This was the main reason why his theory of method may be described as a way of generating *skeptical knowledge*, which is the chosen title of this book. However, Francis Bacon's identification with Socrates and his method with the Socratic method is not a common opinion among specialists who study his writing. We must begin therefore with the question whether he identified with Socrates, and which method of Socrates he took up as his own (Bacon 2000/1605).

He also borrowed from a different from of skepticism that Plato proposed after considering difficulties in the Socratic method. Plato's new skepticism may be called fallibilism, but that does not do his method justice. Unlike Socrates, and other philosophers prior to him, he subjected all his best theories to the harshest and most telling criticism,

J. Hattiangadi, *Francis Bacon's Skeptical Recipes for New Knowledge*,
https://doi.org/10.1007/978-3-031-52585-8_2

which may be called active fallibilism. In this practice he not only acknowledged his fallibility but was actively trying to find out his error. This astonishing Platonic habit was adopted as the overall method of science as Francis Bacon recommended it (Bacon, I.67).

Francis Bacon was also influenced by a third form of skepticism that called itself *Pyrrhonian*, which arose to avoid a difficulty in Plato's active fallibilism. Pyrrhonians described a way of life called "living by the appearances," which is a very conservative but practical way of conducting oneself in a profession. Most Pyrrhonians in Hellenistic times practiced medicine, and their professional practice—their form of skepticism, if you will—included scoffing at causal inquiries and hypotheses and limiting their practice to prescribing well established remedies to patients for treating their symptoms. Living by the appearances is practical, no doubt. Francis Bacon saw that the works covered by a practice could be augmented if we could go beyond the appearances to an underlying reality. Breaking through the appearances to that reality could improve the practice. It would give us new powers over nature, in a way analogous to Pyrrhonian practices that wielded old and established powers over nature. Gaining new powers over nature became a key feature of his method of induction. These two features still distinguish it from any earlier or subsequent theory of method that has been proposed.

However, we must proceed slowly in our exploration of the new philosophy, examining each step closely, because the ideas are so unusual that they are easy to misunderstand. In fact, they have been widely misunderstood, and those who misunderstand will accuse us of error. Plato's active fallibilism, and the way in which power over nature arise very naturally from the exercise of Francis Bacon's method, but they will have to be introduced in later chapters, and all three skeptical influences brought to bear on his method at once still later. Our first task now is to examine whether Francis Bacon identified with Socrates, and our second to clarify the Socratic method (the *elenchus*) that Francis Bacon made his own.

II

When Francis Bacon died in 1627, he was in disgrace. He had risen to the height of power in the English court of James I. His career as a public servant was luminous at its apex, reaching the level of a Lord Chancellor. In 1621, a Parliamentary committee charged him with twenty-three counts of bribery and improper administrative conduct. He would not defend himself publicly against the charge of taking bribes in a trial. Privately, he denied each of the parliamentary accusations, claiming purity of heart.

They remembered him for his power and subsequent disgrace. But the philosophers assembled to study natural philosophy three decades after his death honored his memory as the author of the *New Organon*, as part of his grand project for the *Great Instauration*. His power did not affect them anymore, and the sting of his sudden loss of power was merely a source of amusement and gossip. But they gossiped about an issue that tells us something about his philosophy. It is for this reason that we need to look into some of the salacious details.

The truth of what a philosopher proposes, when the philosophy is new, does not depend on his or her character. To think so is a common, but well-known fallacy. We cannot judge the truth of a proposition until we know what it is. For this we may need to study a philosopher's background. We will go into some salacious details of his life therefore, to find out what he claimed, but not to judge the truth of his claims. Francis Bacon's main philosophical message has been obscure. If nothing else, it has been controversial. Gossip may give us a clue as to what he could have meant, though once we understand it, we must forget the gossip to judge his philosophy.

In the last two hundred years, commentators have dismissed Francis Bacon's *New Organon* as flawed, and his method decried as impossible to practice. At worst, we dismiss him as a charlatan, and at best, as a prophet who turned out to be right—though we may suppose that he prophesied like an oracle without personal knowledge. Earlier, this was not his reputation. Late in the seventeenth century some very successful philosophers honored him for his new method. Those famous natural philosophers, nowadays known as natural scientists, claimed that he had inspired them, and even that they had applied his ideas in their best achievements. These two opinions are very different, of course.

From this difference of opinion concerning his method we know that we do not fully understand his message. For if we did, we would know why he was considered an inspiration in the early Royal Society, but a distraction today. It is too easy to deny one or the other to suit our prejudice. We must look for clues to gain more insight into his thought. If we look wherever we can find it, we are rewarded when we turn to the salacious gossip about him, which gives us a small window into his inner thinking.

III

John Aubrey, in his *Brief Lives*, describes Francis Bacon thus: "He was a παιδεραστής [a pederast]. His Ganimeds and Favourites tooke bribes; but his Lordship always gave Judgement *secundum aequum et bonum* [according as was just and good]. His Decrees in Chancery stand firme, i.e. there are fewer of his decrees r verst than of any other Chancellor." (Aubrey 1898, vol. 1, 71). The gossip: Francis Bacon did not take bribes, but he pleaded guilty to avoid a more severe accusation of masculine love, as then known.

Francis Bacon was a member of one of a handful of families very close to the throne in England, under Queen Elizabeth I. He began life as the son of Nicholas Bacon, the Lord Keeper of the Seal. His mother, a second wife, was a personal Royal attendant to the Queen. He got to know Queen Elizabeth personally as a young boy. He was precocious and much appreciated in her court. He may be described as a well-connected young man when he returned from a spot of diplomatic service in France.[1]

As a seventeen-year-old young man, he went to Paris, in France, with Ambassador, Sir Amias Paulet, to pursue further study after his Cambridge degree. He returned to England two years later when, unexpectedly, his father died. Unlike the children of his first wife, the two young boys by his second wife were left without an estate. Young Francis Bacon then became a lawyer, which allowed him to earn his livelihood. He was elected a member of Parliament in 1584, when he was twenty-three years old. He immersed himself in public life, but struggled at first. He did not do so well financially while Elizabeth was queen. Later he did extremely well for himself, until his disgrace. From his correspondence it appears that he had his idea of a new method quite early, while he was still a struggling lawyer.

When James VI of Scotland became James I of England and united the two realms, there was some friction between King James and Parliament concerning the how to share power. Francis Bacon took up an increasingly prominent part in the dispute between King and Parliament over the decades. He worked hard to smooth over difficulties that regularly arose between the King and Parliament. He soon rose to the rank of Solicitor General, then Attorney General, then the Lord Keeper of the Seal (his father's post, at one time) on his way to becoming the Lord Chancellor of England. He was even briefly an acting regent during King James' brief absence when away in Scotland. Francis Bacon thus was as powerful and influential and well known a politician as any in England when he was at the height of his political career.

At the very height of his power, however, he was accused of taking bribes in office. He did not defend himself. He soon pleaded guilty to the charge, which has surprised many commentators then and since. He swiftly lost power and was disgraced. He died in official disgrace, though he did not spend more than three days in the Tower for his transgressions. He did write and publish some works that were well known to members of the Royal Society. He was thus a very well-known person in England, but notoriously so.

Francis Bacon's life, as well as his trial, was a matter of public knowledge. If he did not take any bribes, and if his "Ganymeds and Favourites" did, then his honesty may be left intact as a public servant. But the gossip is not a defense of his personal character. It does not defend his ability as a prudent head of a household. It is not unlikely that the gossip that John Aubrey reports was widely known and discussed, at least among the ladies and gentlemen of the realm. Aubrey is telling us a story of a tragic fall. Francis Bacon could not admit to his pederasty, more severely punishable than a lesser charge of bribery.

We could suppose that when John Aubrey describes Francis Bacon as a *pederast* he has in mind not merely someone who loves young boys, as a participant in "masculine love" as it was then called. There is, of course, evidence that masculine love was a phenomenon mentioned in English writing early in the seventeenth century. There is also evidence that prominent individuals in England took part in it, and some say that this may have included King James himself.

In the early part of the seventeenth century, carnal experience with another man was punishable by death. It was more serious a crime than accepting bribes. If his "Ganymeds and Favourites" were taking bribes, though he was not, this is a tragic story, because, of course, even if he was completely honest in his dealings as a judge, he could not defend his own honesty by letting it be known that his secret love life had made his household vulnerable to control by his men servants.

In fact, when we look at the facts of the case, we find it shocking that Francis Bacon made no attempt to justify his actions or ask for a fair trial concerning the charge of bribery. Taking bribes, or receiving gifts from litigants, as Francis Bacon said in private, was quite a common practice. John Aubrey hints at this. Francis Bacon's accusers included Sir Edward Coke, a bitter and life long rival who was also Attorney General before Francis Bacon took over his position. When Edward Coke began his legal career, he had no more than £300 in capital, earning perhaps as little as £40 per annum for his living. So reports John Aubrey. From this modest beginning, Edward Coke amassed enough wealth to have a living of eleven thousand pounds per annum as his estate. John Aubrey merely mentions this fact, but discreetly adds nothing about how this estate was accumulated.

Francis Bacon, contrary to reputation, was perhaps an honest judge. He was adamant, in private, even calling upon God to be his witness, that he was pure of heart. He did not adjudicate any cases in which he was accused of receiving a bribe from a favored litigant. The case for bribery brought against Francis Bacon, which led to his disgrace, was brought forward by a man complaining that having accepted his bribe, the Lord Chancellor then did not rule in the complainant's favor. This accusation fits the defense, if true, that Francis Bacon's servants took bribes in his name while he was obliviously conducting trials on their merits.

But the tragedy of Francis Bacon's fall is that the real charge against him—the one being threatened but not made openly—was his pederasty, or masculine love, which would have been abhorrent in those days. In his time there was no age of consent. Whereas we may be inclined to frown upon child abuse, and regard masculine love as a personal choice that needs to be respected, in his day, masculine love was punishable by death, but the age of a partner was not an issue. The bribery charge must be akin to a "plea bargain," ensuring that the committee would not accuse him of common buggery. Such an accusation could have embarrassed both the Earl of Buckingham and King James. They would presumably be tainted by scandal if they were seen as lax about masculine love.

Francis Bacon's punishment appeared to be severe, but it was, in fact, kind. He had to pay a fine of £40,000 to the king, which took precedence over all his other debt. But since the king did not make him pay the debt, he could not be harassed by any other creditors, who were many. Francis Bacon had a lavish lifestyle and was forever on the verge of insolvency. He entered prison in the Tower of London, at His Majesty's Pleasure. The sentence lasted just a few days, as it turned out. But after that, he had to leave London and not come within 12 miles of the court. He could not take part in Parliament or hold public office.

John Aubrey describes Francis Bacon not as a masculine lover but as a pederast, using Greek letters. And this is of some significance. We should note that since there is no age of consent in law in that era, there is no additional moral disapproval of pederasty if we define it as the love of a young man, over and above masculine love. If both include sodomy, it is only sodomy that is disapproved, unlawful, and punishable by death. Pederasty, written in Greek lettering, has more to do with what Aubrey thinks that Bacon saw in the relationships he formed.

In ancient Greece, pederasty was a customary practice within the aristocracy. Typically, an older man would take a younger man as his lover, the latter being a callow youth, or sometimes a man in his late teens. The "erastes," or adult lover would court his "eromenos," his beloved. Young lovers were also called ganymedes and catamites in Francis Bacon's time. The relationship was thus a formally defined and socially accepted practice in ancient Greece. After a period of courtship, the young boy could give over his body, while the older man would become his educator, protector, and role model.

Plato's description of Socrates also displays this characteristic feature of Greek custom. Socrates liked very intelligent young men with a strong sense of intellectual curiosity, who Plato describes as beloved of Socrates. Alcibiades, perhaps, was the most famous. That Socrates was a participant in pederasty has a special meaning for Francis Bacon. It is his fascination with Greece that tells us something, not his sexual preferences. He identified with Socrates.

IV

Bacon's unfinished book of 1602, is strangely entitled *Temporis Partus Masculis*. It is usually translated as *The Masculine Birth of Time*. This is a mysterious name for his new method for a great new science. Several have commented on the mysterious title. Francis Bacon himself writes after his trial and disgrace, in the recently discovered *Abecedarium of Nature*, that by "birth of time" he is referring to the need to take a long time to derive the fruits of his proposed endeavour. But this explanation is made after his disgrace, when he has to be careful not to stress masculine love. He says nothing about his choice of the word "masculine" in the title. What could he say?

Lisa Jardine and Alan Stewart (1998, 117) suggest that it should be translated (freely) as *The Bold Birth of Opportunity*. This title does fit Francis Bacon's writing style very well. But it avoids something else that we should tackle. Why does he use the word "masculus" when he could have used any one of many words for the same purpose? This is not inadvertent. He tells us that he wrote a short essay entitled "Temporis Partus Maximus," or "The Greatest Birth of Time" in 1584. The original work is now lost, so we do not know if it is similar to the manuscript we have. Perhaps it was entirely different. But if we suppose that the title of the lost essay is related to the same manuscript we are considering, then he changed the word "greatest" to the word "masculine." This is hardly a careless choice of a word. One is self-congratulatory, and the other, presumably, descriptive. Francis Bacon was very adept at choosing his words well. He was a great stylist of the English language. What was he getting at?

If he used the word "masculine" to indicate that only men will be working in science in the future, he shows a surprising gender bias. It is surprising because he was the son of a very accomplished woman in his mother, Lady Anne (Cooke) Bacon, a published author. In 1602 Francis Bacon had not published yet. Moreover, there is nothing in his method, whether as he describes it or practiced later, that makes it particularly gender specific. Women have done well enough at philosophy, if we are willing to pay attention to their writing, that they would challenge the gender bias that is common in his time, and in ours.

Evelyn Fox Keller (1985) has pointed out that Francis Bacon reveals great gender bias altogether in his use of the English language. This is true no matter how we interpret the curious title of this early precursor to the *Novum Organum*. His is, moreover, a gratuitous and unnecessary bias to impose upon science as he thinks we have to practice it. We need to ask how he could propose a new method for science and what he was proposing as a "masculine birth." If we wish to defend his gender bias or his benign character, it would be a difficult task. Our task here is different. We must study his words to find out the nature of the experimental method that he proposed as part of the "masculine birth" that he is announcing. Even if we judge its origin reprehensible, we need not therefore neglect what he meant.

The fact that John Aubrey uses the Greek word "pederast" is a clue. Francis Bacon may be thinking of Greek philosophy. Masculine birth is then a substantial clue. Perhaps we should interpret *The Masculine Birth of Time* by referring it back to Plato's *Theaetetus*. Socrates describes himself in this dialogue as akin to a midwife, but he helps another person (in the dialogue, *Theaetetus*) give birth to an idea. Just as women give birth to a child, so men to an idea, and he (Socrates) helps deliver them.

Socrates and Theaetetus are trying to find a definition of knowledge:

> **Soc**. … [T]ry and bring the many sorts of knowledge under one definition.
>
> **Theaet**. I can assure you, Socrates, that I have tried very often, when the report of questions asked by you was brought to me; but I can neither persuade myself that I have a satisfactory answer to give, nor hear of any one who answers as you would have him; and I cannot shake off a feeling of anxiety.

Soc. These are the pangs of labour, my dear Theaetetus; you have something within you which you are bringing to the birth.
Theaet. I do not know, Socrates; I only say what I feel.
Soc. And have you never heard, simpleton, that I am the son of a midwife, brave and burly, whose name was Phaenarete?
Theaet. Yes, I have.
Soc. And that I myself practise midwifery?
Theaet. No, never.
Soc. Let me tell you that I do though, my friend: but you must not reveal the secret, as the world in general have not found me out; and therefore they only say of me, that I am the strangest of mortals and drive men to their wits' end. Did you ever hear that too?
Theaet. Yes.
Soc. Shall I tell you the reason?
Theaet. By all means.
Soc. Bear in mind the whole business of the mid-wives, and then you will see my meaning better:-No woman, as you are probably aware, who is still able to conceive and bear, attends other women, but only those who are past bearing.
Theaet. Yes; I know.
Soc. The reason of this is said to be that Artemis-the goddess of childbirth-is not a mother, and she honours those who are like herself; but she could not allow the barren to be mid-wives, because human nature cannot know the mystery of an art without experience; and therefore she assigned this office to those who are too old to bear.
Theaet. I dare say.
Soc. And I dare say too, or rather I am absolutely certain, that the mid-wives know better than others who is pregnant and who is not?
Theaet. Very true.
Soc. And by the use of potions and incantations they are able to arouse the pangs and to soothe them at will; they can make those bear who have a difficulty in bearing, and if they think fit they can smother the embryo in the womb.
Theaet. They can.
Soc. Did you ever remark that they are also most cunning matchmakers, and have a thorough knowledge of what unions are likely to produce a brave brood?
Theaet. No, never.

Soc. Then let me tell you that this is their greatest pride, more than cutting the umbilical cord. And if you reflect, you will see that the same art which cultivates and gathers in the fruits of the earth, will be most likely to know in what soils the several plants or seeds should be deposited.
Theaet. Yes, the same art.
Soc. And do you suppose that with women the case is otherwise?
Theaet. I should think not.
Soc. Certainly not; but mid-wives are respectable women who have a character to lose, and they avoid this department of their profession, because they are afraid of being called procuresses, which is a name given to those who join together man and woman in an unlawful and unscientific way; and yet the true midwife is also the true and only matchmaker.
Theaet. Clearly.
Soc. Such are the mid-wives, whose task is a very important one but not so important as mine; for women do not bring into the world at one time real children, and at another time counterfeits which are with difficulty distinguished from them; if they did, then the, discernment of the true and false birth would be the crowning achievement of the art of midwifery-you would think so?
Theaet. Indeed I should.
Soc. Well, my art of midwifery is in most respects like theirs; but differs, in that I attend men and not women; and look after their souls when they are in labour, and not after their bodies: and the triumph of my art is in thoroughly examining whether the thought which the mind of the young man brings forth is a false idol or a noble and true birth. And like the mid-wives, I am barren, and the reproach which is often made against me, that I ask questions of others and have not the wit to answer them myself, is very just—the reason is, that the God compels me to be a midwife, but does not allow me to bring forth. And therefore I am not myself at all wise, nor have I anything to show which is the invention or birth of my own soul, but those who converse with me profit. Some of them appear dull enough at first, but afterward, as our acquaintance ripens, if the God is gracious to them, they all make astonishing progress; and this in the opinion of others as well as in their own. It is quite dear that they never learned anything from me; the many fine discoveries to which they cling are of their own making. But to me and the God they owe their delivery…Once more, then, Theaetetus, I repeat my old question, "What is knowledge?"-and do not say that you cannot tell; but quit yourself like a man, and by the help of God you will be able to tell." (*Theatetus* 148d–151d)

Francis Bacon saw himself as bearing an idea that Socrates or Plato had sown. He is tending to it as a midwife, as Socrates did in the *Theaetetus*. His method will produce offspring only when others take it up and give birth to new ideas. Nothing could be clearer as a reference to a "masculine birth" of a new idea than Plato's description of it in the *Theaetetus*. "The masculine birth of time" is the new idea that has at last been conceived by Francis Bacon so many centuries after Plato's writing. Moreover, Francis Bacon does not give birth to a new philosophy. Like Socrates he will help others with his new method, and these others will give birth to the new science.

He wrote the two completed chapters of *The Masculine Birth of Time* in the form of a dialogue with a young man, whom the author calls "my son." Francis Bacon shows how the mind is to be freed from past philosophies so that a practical and enlightening science can be obtained by someone who is willing to give himself over to the author. He suggests that the young lad should not emulate the thinkers of the past because they would lead him astray. Who then should he emulate? "Take heart, then, my son, and give yourself to me so that I may restore you to yourself."

Francis Bacon identified with Socrates, it seems. But which Socrates? Plato's character Socrates has been repainted by commentators in many different ways. Two stand out as relevant for us. One may be called the combative persona of Socrates, and the other the contemplative. The former stands out prominently in Plato's *Apology*, in which Socrates explains to the jurors at his trial why he has been misunderstood and resented by the powerful in Athens. The contemplative Socrates is different. He was identified by Aristotle as the first to practise induction, or epagoge. Aristotle has sometimes been understood to mean that Socrates practised induction by simple enumeration.

But the contemplative Socrates that Gregory Vlastos (1994) portrayed in his essay on Plato's *Gorgias* is more subtly interesting. Socrates is shown using his method of the *elenchus*. The method uses criticism and refutation, but Socrates supposed that he could learn affirmatively about the wisdom Socrates sought. Socrates' method fails, as Plato portrays it in the *Gorgias*. The dialogue is a turning point in Plato's philosophical development, as Gregory Vlastos traced his thought, from the method of the inductive *elenchus* to the method of hypothesis, which Plato took up in the *Meno*. We will begin with the more common portrayal of Socrates the combatant.

V

Socrates was born to a family of craftspeople. His mother was a midwife, and his father a stonemason or a potter.

Socrates never wavered in his expressed admiration of artisanal knowledge where they are expert. In every record of his conversations where craftspeople are mentioned, in Xenophon's *Memorabilia* and throughout Plato's dialogues, and in Aristotle's accounts, Socrates shows respect, whenever appropriate, for craftspeople's knowledge. Never once does he doubt that artisans know. As opposed to this, he challenges the wisdom of sophists when he questions them about their wisdom. In contrasting the sophist's lack of wisdom with the expert's knowledge in her field, Socrates shows a shortfall in their sophistical wisdom. We cannot dismiss his deference to artisans as a feeling of loyalty to his parents. It was a firm conviction, which he had thought out.

Being born of and brought up by craftspeople should be understood as a humble background in ancient Greece, as compared, for instance, to the family background of Plato, who was socially more ascendant. Plato could claim remote royal descent, but he admired Socrates, the child of crafts people. Socrates, and all his students, including Plato, acknowledged the wisdom of some experts in their own proper area. That Socratic trait is a vindication of the importance to him of his own background, modest though it may be. Socrates did not concede knowledge by pedigree.

Socrates was born at a special time in Athenian history. Athens had become a great naval power, and a wealthy centre of maritime trade, thus attracting the military interest of even the Persian (Achaemenid) Empire, the mightiest of that time. Attracted by the wealth in Athens, wise men (sophists) from around Greece, Italy, Turkey, Libya, Macedonia and beyond, traveled to Athens to profit from teaching Athenians some newly discovered wisdom. Athens was a wealthy intellectual center. Many sophists came to instruct.

When Socrates was a young boy, Athens also underwent a tremendous egalitarian change when Pericles' party of democrats was in power. Under Periclean democracy, the ordinary men of the land, including the skilled artisans of Athens, were now allowed to vote and take a larger part in public life. They began to feel more empowered than ever before.

When we read Plato's early "Socratic" dialogues, Plato's admiration for Socrates' pugnacious conversations shines through. We know from Plato's writings how remarkable a philosopher Plato was. We can conclude that Socrates must have been very clever and challenging to influence Plato so profoundly. He must have won some arguments with some famous people. It is even more remarkable because Socrates was a son of an artisan and a midwife. Plato himself did not seem to share our admiration for an egalitarian democracy. So his admiration of Socrates must have been for the philosopher himself.

It would be surprising if a plain boy, such as Socrates, did not initially find himself at a disadvantage when conversing with the great wise men or sophists visiting Athens. They came over because they had a reputation for wisdom. They hoped to profit from it. What could Socrates, the son of a potter and a midwife, say to them? He would be a rude intruder in an exalted company who knew about important things about public life. They had come to teach, not to learn.

In a democracy, however, everyone can judge a public policy, as Pericles had allowed. Thus, Socrates was empowered to discuss matters of public importance despite his modest background. Socrates seemed to have found an interesting balance between his humble background among the crafts, and his aristocratic Athenian public life, where he discussed public virtues even with feudal aristocrats.

A wise person, and a teacher, may not know how to shape a jar, but he could still be clever with words. A theory of nature or matter may explain the nature of a pot. But if their words did not mention the sculptor's aim in fashioning it, they would seem ridiculous to a sculptor. What do wordsmiths know? How can their flimsy words compare with the concrete knowledge that guides the hands of an artisan? If you know what a pot really is, then make one and show me. Socrates likely began his education in philosophy with this attitude, adopted from his parents and their friends.

But he will soon find that there are important questions in the larger world to which he had no answers. In a democracy, and especially in a democracy, ordinary people need to understand public policy questions. Blacksmiths may make horseshoes, but who makes the laws? Wordsmiths cannot be wholly ignored in politics, even if they did not have all the answers, as craftspeople of the public world. Socrates challenged the wordsmiths' claims about knowing wisdom. But he also denied knowledge of it. His claim of ignorance is very self-deprecating, on one reading of it. On another, it is also a very aggressive kind of ignorance. Socrates used it as a tool of discomfiture for some interlocutors. He made them seem to claim more than they knew. There is a lively debate about whether he was humble in his ignorance or ironic in his demeanor.

It was likely a little of both. There was likely drama, occasionally, when he outwitted a famous man in an argument. Here was a very local boy, just a son of a local potter and a local midwife, salt of the earth, who was challenging the great wise men from all over the accessible world. His triumphs would appeal to his young Athenian friends and admirers. In a wealthy and burgeoning trading center, and especially in an egalitarian democracy like Periclean Athens, there must have been many young men seeking excitement, for whom Socrates, the local terror, must have been fun to witness.

In Plato's *Apology*, at his trial that led to his execution, Socrates speaks about craftspeople, among others. Socrates was seeking people wiser than himself. He met a great number of important personages, including sophists from distant lands, and local Athenians in high office, and reputable citizens, who were believed to be wise. The combative Socrates found them not to be so: "Last of all, I turned to the skilled craftsmen. I knew quite well that I had practically no technical qualifications myself, and I was sure I would find them full of impressive knowledge…They understood things that I did not, and to that extent they were wiser than I was. But these professional experts seemed to share the same failing that I had noticed in the poets. I mean that on the strength of their technical proficiency they claimed a perfect understanding of every other subject, however important, and I felt that this error more than outweighed their positive wisdom." Socrates is among the craftspeople, friends of his family, appreciating their specialized knowledge. But here he is denying them wisdom. If Socrates was a skeptic, then he was a skeptic who appreciated the partial and inadequate wisdom of the crafts. He admired the crafts. But he was skeptical of the sophist's craft.

VI

Socrates was put to death by his beloved city of Athens. Socrates' influence as a philosopher greatly increased after his death. He was already a local hero among Athenians when he was still alive, as a local craftsperson's son with a ferocious method of questioning. But he did not write down any of his philosophical views.

After his death, a number of writers took up the task of writing a defense of Socrates from the charges laid against him at his trial. Plato's *Apology* is to be found in this genre of writing. In it, Plato reports the speeches Socrates made in his own defense at that famous trial. In these speeches is to be found a summary of Socrates' public activity as a philosopher. Socratic skepticism can be found clearly stated in this remarkable account. We may take Plato's formulation of it in the Apology as a definitive statement of Socratic skepticism.

Xenophon records that the charges laid against Socrates were that he "refused to recognize the gods acknowledged by the state, and invoked strange divinities of his own; he is further guilty of corrupting the young." (Xenophon 1897/4th c. BCE, 1). There would be no reason for Socrates to fear the first charge at all in a modern liberal democracy. The right to freedom of religion or worship would protect him. As for the second charge, had they accused him merely of corrupting the youth's minds, he would be protected by his right to freedom of speech.

In his defense, Socrates avowed ignorance. His ignorance is a uniquely personal form of skepticism. It is also a very humble statement, though some scholars have suggested that he was not sincere. Socrates' professed ignorance was perhaps stated ironically. In the speech reported in Plato's *Apology*, Socrates described a general state of hostility towards him among influential Athenians. This hostility towards him, he speculates, must have been caused by his public activity as a questioning or skeptical philosopher.

He described it as follows. His friend Chaerophon went to the oracle at Delphi and asked if there was anyone wiser than Socrates. To this, the oracle replied that there was no one. Apollo—son of Zeus, and divine by nature—who supposedly spoke through the oracle, does not lie. Socrates must therefore accept that he is the wisest of all humans. But this is odd, indeed, because his characteristic stance as a craftsperson's son was that he was ignorant of the abstract matters that the wandering wise people taught. Going by reputation, many wise Athenians must be considered wiser than he, since he was ignorant. He had paid to learn from some of these great sophists, although the money tendered as his fee was likely a form of charity obtained from his wealthy young friends. Socrates was also an admirer of the wise people of Athens, he claimed. How could he be wiser than they?

Socrates' decision upon hearing the oracular saying is well known; he decided to test the statement; he went around and asked important and influential people around town some difficult questions about their reputed wisdom. He asked them what they knew; and when they found out after being cross-examined by the clever Socrates that they did not know what it took to be wise, they often became angry. Socrates seemed to be making fools of them in a public place, where all the assembled youth could laugh at their discomfiture. So resentment was building up, he claimed. Socrates persevered because he had to determine what the oracle meant in calling him the wisest. He went around and made a great many more important Athenians angry. Hence, he says, there is a lot of hostility and resentment towards Socrates among those with high reputations in Athens.

Socrates' resolution of the paradox is what interests us. The paradox is posed by Socrates' own avowed ignorance when conjoined with Apollo's oracular description of Socrates as the wisest among all men. He resolved it thus: Though he knew nothing, at least Socrates knew that he knew nothing; all the others thought that they knew, but they did not know; they were unaware even of their own ignorance. He is wiser than they were in this, and this alone, that he at least knew his own ignorance.

VII

Francis Bacon took his inspiration from Socrates for his new method of interpreting nature, or of induction. He learnt a skeptical method of gaining knowledge from a contemplative Socrates, who argued skeptically to affirm ethical doctrines. The persona of this Socrates is subtly different from the combative self-portrait Socrates painted in his speech of the *Apology*. At his trial, he reported refuting others, and thereby offending some important people of Athens. He claimed he did not know. Yet he was also a contemplative philosopher who propounded some startling ethical doctrines.

A contemplative Socrates found in Plato's early dialogues claimed that virtue (or excellence at something good) is knowledge. It remains a surprising claim, because if we reflect upon and compare the concepts, they seem very different to us. Another claim that must also have startled contemporary Athenians is about justice. He claimed that it is better for anyone to suffer an unjust harm than to inflict it unjustly on another. In Athens at that time getting justice was considered getting even with someone. Both the ethical claims were quite contrary to customary understanding.

Socrates only discoursed about ethical and political matters, of course. The interesting thing about this contemplative persona of Socrates is that he argued affirmatively for surprising ethical doctrines by employing only the *elenchus*, which is primarily a method of criticism and refutation, as Gregory Vlastos pointed out in some very elegant philosophical exegesis. The combative Socrates of the *Apology* used the same kind of cross questioning, while concluding that he knew nothing, he said. Making affirmative claims to know about justice while denying that he knew anything amounts to a self-contradiction, or a paradox. If he did not know, how could he assert such strange doctrines as knowledge? It would seem that Plato studied the paradox, and concluded that the Socratic *elenchus* demanded too much of us to acquire affirmative knowledge. Francis Bacon, however, solved the problem, exemplified in his new method of science.

Gorgias of Leontini was an interesting orator and philosopher, one of the early and influential sophists, whose main philosophical writings have been lost. He is remembered for posing many paradoxes, and also for producing brilliant rhetorical techniques which have been largely preserved for posterity. Of his philosophy, we remember only a gist, in which he appears to us primarily as a critic of Parmenidean realism. He argued that *there is no reality (or being). If there were any reality, we could not grasp what it is. If we were to grasp it, we could not express it to others. And if we were to express it to others, they would not understand it* (Sprague 1972, 42).

Gorgias is often described as a nihilist, but that is a dubious interpretation of this text because he admits that there is really a mind, and it could grasp reality, were there only something beyond to grasp. Language and language speakers also must be real. We can understand his gisted philosophy better as a quasi-legal brief for relativism, or the doctrine that there is no reality beyond us humans. Like Protagoras, he taught, as contemporary relativists may say, that we grasp, speak, and understand as real merely those imaginaries which we construct or create as real among societies of human beings. The quasi-legal formulation of Gorgias' main argument is also not surprising. Famous sophists of that time traveled to Athens to teach their new students the rhetorical art of defending themselves against a legal accusation or to prosecute one at a trial. For a relativist, the point of argumentation is to convince others, and a legal trial is an important venue for it.

There is a certain religious quality to Socratic ethical theories. That it is better to be harmed than do harm sounds like the sort of thing one would hear from Gautama Siddhartha or Jesus of Nazareth. If we read him as a prophet, we can read his disavowal of knowledge as a rhetorical or pedagogic device in the service of a new Socratic religion. Since he knew this deep truth, his avowed skepticism must be a deception, a form of irony. Some continue read Plato and Plato's Socrates as ironists in this prophetic vein. They resolve the skeptical paradox by saying he was not a skeptic, after all. Since our interest is in skepticism, we must take the contemplative Socrates literally, paradox and all.

Plato could not resolve the paradox. He gave up on showing how a Socratic *elenchus* miraculously generates affirmative knowledge. He went on in the *Meno* to describe and then adopt a different method, which was widely used by mathematicians, named the method of hypothesis. His method of evaluating hypotheses is very similar to the *elenchus*, for it relies on refutations. But it remains a method of refutation and criticism alone. In the more recent words of Karl Popper, Plato fell back on critical rationalism, eschewing induction. This transition is important for Francis Bacon, because he called the Socratic method *the method of interpretating nature*. He recommended it. He called hypotheses *anticipations of the mind*, which has been widely used before and after Plato, which he did not recommend for gaining knowledge.

The *Gorgias* is a watershed dialogue, in which Plato shows the full force of the Socratic claim to have affirmative knowledge of justice. His *elenchus* is on full display in his exchange with Gorgias and Polus. Gorgias offers a piece of wisdom; and he is asked if he believes some ancillary doctrines but when he admits it, Socrates shows that Gorgias ended up with a self-contradictory consequence. Polus offers Gorgias a way out by denying the ancillary doctrine, but must admit that Socrates is right after his contrary ancillary belief was also refuted—in a system that included yet another ancillary belief of Polus'. This shows the power of the Socratic *elenchus* in generating refutations. But what about affirmative knowledge?

Callicles comes forward and says that Polus gave in too easily. He could have chosen to defend his contrary position by a different manoeuvre. It is never enough to refute a piece of wisdom, because there could more like Callicles who find different manoeuvres to deny any Socratic wisdom by deployjng them.

Gregory Vlastos pinpoints this dialogue as Plato's turning point away from an *elenchus* that can teach affirmatively. In this dialogue it becomes clear to Plato, to Gregory Vlastos and to us, that Socrates makes too bold a claim to know based on the method of refutation—that if he claimed that his theory of justice is true then he can elicit its knowledge in his interlocutor by using the *elenchus*. To argue for his own theory of justice as knowledge, he had to show that *every contrary belief system is self contradictory*. That is too strong a claim to be shown. Plato abandoned the method of cross-examination characteristic of his early dialogues in the next few dialogues, until writing the *Meno*.

It is important to look at the difference between the method of hypothesis that Plato adopted and the Socratic method of elenchus that he abandoned, for in many respects they were similar. In each, some philosophical work is done by criticism and refutation. In each the end result is usually and typically a denial of the knowledge sought. A bold hypothesis may be affirmed and examined, but it cannot be endorsed as knowledge. It may only be used to shed light on what follows logically from it. The prominent differences between the methods are two. When it is successful, (or occasionally, or if at all) the first method can give us an affirmative principle as knowledge. The method of hypothesis can make no such claim. Another difference is that the Socratic *elenchus* is restricted to cross-examining personally held beliefs, while the method of hypothesis is free from this constraint.

The importance of believing or trusting the truth of what is being examined is an important feature of what Socrates asks of his interlocutors. He insisted on this, and he would not examine opinions of what may be true, or hypotheses that are purely conjectural. In fact, Socrates seems characteristically to engage in discussions with people about whatever is very close to them, a part of their self image. For instance, he engages with Laches, a general, about courage. A general will have given a thought to courage and he must have strongly held opinions on it. That is precisely where he can work with the beliefs of a general. Somehow, Plato portrays Socrates as engaged in cross-examining strongly held personal beliefs of his interlocutors.

The interlocutor is like an instrument, we may say, who is used by Socrates to investigate the interlocutor's beliefs. Galileo Galilei peered through a telescope and found the stars in the milky way. They were not known earlier because without the instrument, they were not evident to the naked eye. So too, with Socrates, who found ethical knowledge by cross-examination of an instrument, another human being, who arrives at an affirmative principle due to the cross-examination. How anyone arrives at a new principle that is knowledge by this method is a much more difficult thing to explain. Plato, it seems, could not explain it. After the *Meno*, he embarked on a wonderful lifelong journey of examining bold conjectures, not only in ethics and politics, but in almost every subjects then known to philosophers. He continued to revere the Socratic method. He describes it often as the highest from of understanding, even though he could not show how a personal *elenchus* can yield wisdom affirmatively.

Francis Bacon solved the Socratic problem, going beyond Plato or Aristotle or other philosophers who practised the method of anticipating nature. His modified Socratic *elenchus* showed us how to gain affirmative knowledge of principles.

Note

1. Biographical data about Francis Bacon are drawn from Jardine and Stewart (1998), as well as Spedding (1878) and Mallet (1768).

References

Aubrey, John. 1898. *Brief Lives, Chiefly of Contemporaries, Set Down by John Aubrey, between the Years 1669 & 1696*, 2 vols. Ed. Andrew Clark. Oxford: Clarendon Press.

Bacon, Francis. (1605) 2000. The Advancement of Learning. In *The Advancement of Learning*, ed. Michael Kiernan. Vol. IV of *The Oxford Francis Bacon*. Oxford: Clarendon Press.

Jardine, Lisa, and Alan Stewart. 1998. *Hostage to Fortune: The Troubled Life of Francis Bacon*. New York: Hill & Wang.

Keller, Evelyn Fox. 1985. *Reflections on Gender and Science*. New Haven, CT: Yale University Press.

Mallet, David. 1768. *The Life of Francis Bacon, Lord Chancellor of England*. London: Printed for A. Millar.

Spedding, James, ed. 1878. *An Account of the Life and Times of Francis Bacon, Extracted from the Edition of His Occasional Works*. London: Trubner.

Sprague, Rosamund Kent. 1972. *The Older Sophists*. Columbia: University of South Carolina.

Vlastos, Gregory. 1994. The Socratic Elenchus: Method Is All. In *Socratic Studies*, ed. Myles Burnyeat, 1–38. Cambridge: Cambridge University Press.

Xenophon. 1897. *Memorabilia*. In Vol. 3 of *The Works of Xenophon*, ed. H.G. Dakyns. London: Macmillan & Co.

3

The Active Fallibilism of a Situational Skeptic

I

Francis Bacon's greatest skeptical advance was to harness the Socratic e*lenchus* to learn affirmatively about the principles of natural philosophy. Francis Bacon's form of the *elenchus* was new in several ways, as we will see when we study it as his scientific method. Unlike the Socratic *elenchus*, modern science is not a personal or individual accomplishment but a communal effort. He imagined it as a government department, or a Royally funded institute. Still, when a community eventually took it up, its funds were provided not by the throne but by some independent scholars who formed the Royal Society of London. The Society was sufficiently distinguished that a detailed history was written of it five years after its formation (Sprat 1667).

Another difference from Socrates concerns the medium of communication. Socrates did not write a single sentence for posterity. He conversed, and others noted his conversations and wrote about them afterward. Plato, the most famous raconteur, even opined in a late dialogue that writing was inferior to oral discussion in philosophy. Francis Bacon's *elenchus*, in contrast, made writing a record in a journal an indispensable component of the method. The results had to be written in a particular way to be useful.

J. Hattiangadi, *Francis Bacon's Skeptical Recipes for New Knowledge*,
https://doi.org/10.1007/978-3-031-52585-8_3

We will also see that Francis Bacon restricted his elenctic examination to natural science. He wanted to exclude and protect certain ethical and theological doctrines from skeptical scrutiny. Therefore, modern science derived shape from its Christian background, particularly its Anglican background, in which some moral and political laws are of divine origin.[1]

We need to explore a fourth difference between the elenctic methods. The Socratic elenchus, though skeptical, may have been conceived to produce affirmative results. In the *Gorgias*, Socrates claimed to *prove* an ethical doctrine. He claims later in the same dialogue, as Gregory Vlastos (1994) pointed out, that "arguments of iron and adamant" have clamped down his moral claims (*Gorgias* 508e). But an *elenchus* can only reveal inconsistencies in opinions, which marks their falsity. We can celebrate logical consistency wherever we find it, of course, but such consistency is a mark of truth only if it is unique among consistent beliefs. Every contrary opinion must be refuted. In the language of latter-day skeptics, Socrates was being dogmatic. He presumed uniqueness for the consistency of his ethical doctrine, but perhaps he had no reason to assume it. In Gregory Vlastos' celebrated reading, Plato recognized that this Socratic claim is too strong. He abandoned the Socratic *elench*us in three dialogues following the *Gorgias*. When he took it up in the fourth, the *Meno*, he adopted the method of hypothesis.

Francis Bacon recommended that modern science adopt a form of skepticism that today is called fallibilism, or the recognition that any of our beliefs may be mistaken. He adopted fallibilism in its Socratic or Platonic forms. His fallibilism was *active*. Not only did he allow that any of his hypotheses may be mistaken, even his most trusted belief, but he actively pursued their refutation. His Socratic self-examination throughout his life was unsparing. Francis Bacon recommended this form of active fallibilism to the community of natural philosophers. Being an active personal fallibilist is not easy. Not many philosophers can do it. But Baconian science is a group activity. It easier in a community where the protagonist and critic are different.

Science is like a pyramid, wrote Francis Bacon. Ascending from the lower levels, we keep applying the same *elenchus* as we climb higher to ever more general principles. Because misleadingly he called the ascent "induction," the *elenctic* process can be missed or misunderstood. It is in the upward or inductive direction—but we must be prepared to examine, relinquish and modify whatever middle axiom we discover. Active fallibilism was an extraordinary Platonic (or Socratic) contribution to philosophy. Francis Bacon merely endorsed it as a norm for collecting ever more knowledge.

II

The *Meno* depicts a handsome young man, Meno, a follower of the Sicilian sophist, Gorgias of Leontini, who is asking Socrates some very uncomfortable questions. The *Meno* was written after the *Gorgias*, in which it becomes clear that Socratic refutation is inadequate to prove an ethical principle. In the *Meno*, Socrates encourages Meno to challenge him with difficult questions. Indirectly, perhaps, Plato is conversing with the relativist, Gorgias, who is in the background, raising difficulties for Socrates. Gorgias was well known in antiquity to have been generous in suggesting paradoxes to his opponents, and he may have posed this one, too. We have lost most of the writings of Gorgias, so we cannot say. Socrates plays a subtly different role here than customary in Plato's earlier dialogues.

Socrates is studying whether virtue is teachable, a question he often poses to sophists, who were self-proclaimed teachers of virtue. He and Meno are in perplexity. It is a snappy dialogue, as if Plato has taken up the *elenchus* again. Plato's clever device of structuring the early dialogues in this way captures Socrates' profession of ignorance beautifully, combined with his brilliance at refuting of his interlocutor's asserted views. This structure of an early dialogue is a rhetorical device. Conversations do not begin and end in this formal way. Plato devised the form to capture the kind of sharp questioning and passionate moral attitude that Socrates exhibited. No actual conversations may have begun and ended as portrayed, but in their portrayal, Plato captured something special about Socrates. They are framed within the discussion a question that is often not answered. The frequent aporetic endings also say something special about Socrates, highlighting his willingness to say, "I do not know." It would be much less dramatic to describe this feature in another way. But a part of the dialogue is different: the learning in this case is not all Meno's. In this it is not like some earlier Socratic dialogues. The Meno shows Socrates or Plato learning something, which is expressed in the voice of Socrates. Here, in the *Meno*, a student of the redoubtable sophist, Gorgias, asks Socrates a difficult question. He has come from Gorgias' school in Larissa to Athens to beard the lion in his den. The usual dialogic roles are reversed.

We must remember that Meno was from the school of Gorgias, in Larissa, not far from Athens. As a relativist, he could not accept Socratic skepticism over the issue of knowledge. His relativism is not explicit in the *Meno*, but we can see it in context. Meno raises the issue about the Socratic claim that he seeks knowledge because even if there were a reality to know, as Gorgias might have said, we could not know it. Meno does not propound relativism as a doctrine but merely asks a question to undermine Socratic skepticism. To undermine skepticism is to promote relativism. This dialogue is therefore no longer Socratic in an old-fashioned sense. Plato has become an active fallibilist already, allowing the Socratic *elenchus* to be questioned—by a sophist. He adopts the motto of his teacher, Socrates—he sets out to examine himself. He is examining the Socratic *elenchus*.

> Socrates: .... I am ready to carry out, together with you, a joint investigation into what it [knowledge] is....
>
> Meno: But how will you look for something when you don't in the least know what it is? How on earth are you going to set up something you don't know as the object of your search? To put it another way, even if you came right up against it, how will you know that what you have found is the thing that you did not know?
>
> Socrates: I know what you mean. Do you realize that what you bring up is the trick argument ("the hoary old riddle") that a man cannot try to discover either what he knows or what he does not know? He would not seek what he knows, for since he knows it there is no need of the inquiry, nor what he does not know, for in that case he does not even know what he is to look for. (*Meno* 80d-e)

Meno, the young sophist, posed this riddle to Socrates, and to all skeptics and realists in posterity. What follows is the beginning Plato's characteristic philosophy.

> Meno: Well, do you think that this is a good argument?
> Socrates: No.
> Meno: Can you explain how it fails?
> Socrates: I can. I have heard from men and women who understand the truths of religion …. (*Meno* 81a)

His refers to a religious doctrine. But he does not mention which religion, because it does not matter. The idea is an impersonal one, a prelude to his new philosophy. It is not a Socratic doctrine since Plato learned it from someone else—"men and women who understand the truths of religion." The principle he introduced has since become well-known as a central piece in Plato's philosophy of his middle period: knowledge is gained by recollecting something that already exists in our souls. It is a hypothesis.

This passage resolves the dilemma Meno had set up in criticism of Socrates' effort to inquire into knowledge. Plato suggests we do not know the answer we seek until we *recollect* what it is. Not knowing it, we seek it, But in another sense of the word "to know," we know what it is if we can recollect it. It is already in our souls to be recollected. We know it now because we recognize some resemblance to something innate in the soul. As we advance from the *Meno*, Plato's thought attached increased importance to the objects hidden in the soul. Plato's Socrates, in this process, becomes a mouthpiece for Plato's ideas. Since Plato is defending Socrates here, the extension does not seem illegitimate. It is as if he shows what Socrates could have said, or perhaps should have, in defense of his life-long search for wisdom.

He illustrates recollected knowledge by conversing on geometry with a slave boy randomly chosen from Meno's entourage. The little dialogue with the boy is also a celebrated passage within the larger dialogue. The choice of the boy is appropriate. An enslaved person would not have a formal education and so would not have advanced mathematical knowledge. Moreover, Socrates could not have coached a boy just visiting Athens with Meno's entourage.

This dialogue illustrates how a mathematically uneducated young boy can work out the correct answer to a mathematical problem for himself when questioned adroitly by Socrates. By answering Socratic questions, he could find some truths on his own—which happened to be some advanced truths of the geometry of the day. Moreover, the under-educated boy does so solely by what he can judge for himself to be true. The use of the *elenchus* requires it.

This famous interchange also follows an interrogative path like some early dialogues of Socrates. This mini-dialogue begins with a geometrical question to which the boy answers incorrectly. Socrates refutes this answer using only claims to which the boy will assent. The refutation first numbs the boy—like the sting of a stingray. He does not know how to proceed. Then, the boy is taken through a process of thinking about the geometrical issues. Socrates follows up with a series of questions that elicit answers from the boy that the boy can judge as true. Plato finally shows that the boy knows the answer: he shows that the uneducated boy understands an advanced geometrical discovery of his day. He initially gave the wrong answer, yet he knew the true answer because he assented to it later by relying on his personal judgment.

This example is rare, in which a Socratic *elenchus* leads directly to knowledge. The interlocutor successfully gained knowledge, but not Socrates, of course, since he knows nothing. An *elenchus* is successful at generating affirmative knowledge. In this mini-dialogue (within the dialogue, *Meno*), we do not conclude by reaffirming our geometric perplexity. In this respect, the argument illustrates how an affirmative ending to a search is possible, as Socrates asserted in the *Gorgias*. However, it is in mathematics that he finds it working. For that reason, perhaps Plato became interested in the esoteric subject. Another reason may have been because by refuting a claim, they would establish its negation as a theorem. There is a fascinating similarity, but not an identity, between a proof by *reductio ad absurdum*, as it was later called, and the Socratic *elenchus* that promised affirmative knowledge of moral principle.

When we began considering Meno's dilemma, we read that there could not be any middle option: either we know or do not know what we seek. We should have stopped right there. Plato suggests that the solution to the dilemma is to reject that claim. We can avoid having to choose between the horns of the dilemma. They are simultaneously true; in one sense, we know what we seek, and we have known it all along, though, in another sense of the word, we do not possess the knowledge at the tip of one's tongue and, not knowing, therefore continue to seek it.

With this turn of the argument in the *Meno*, we note several themes associated with Plato's philosophy. The doctrine of the complex soul is here: one part knows something that another does not. The principle that the soul forgets its knowledge when a person is born, during birth pangs as it were, is also proposed by Plato, suggesting that the soul migrates from body to body and thus outlasts its death. Perhaps the soul is immortal, unlike the body, though that point is only argued later in the *Phaedo*. We also wonder how knowledge is such that it can be lodged in the soul in this manner. It remains unexplained in the dialogue.

In the *Phaedo*, Plato connects the doctrine of recollection with a hypothesis concerning the separate reality of Forms or Ideas. They fit the Socratic style of questioning, which asks interlocutors what a virtue is, in itself. In the *Phaedo*, Plato shows how the Forms can answer to the kind of reality that Parmenides was seeking, an unchanging something that underlies change. One of the central doctrines associated with Plato's thought is thereby introduced. Our concern here, however, is a noted feature of this dialogue. The *Meno* is an example of Plato's form of skepticism, which may be called "active fallibilism." Here he is challenging the dialectic that Socrates and he practiced. To defend that dialectical form, he invents the hypotheses of recollection, eternal and divided souls, Ideas or Forms, causal accounts of things, and later of a God who knows, i.e., who can put these Ideas in our soul. It may seem extravagant to invent all that to make sense of the Socratic quest for knowledge. However, it illustrates how active fallibilism can help us advance to new hypotheses.

III

Our quest is to understand Plato's form of skepticism, namely active fallibilism, because it influenced Francis Bacon. We must look at a somewhat neglected aspect of the *Meno* and the problem of knowledge proposed in it.

When we study the *Meno*, we are dazzled by the extraordinary doctrines proposed by Plato. We see Plato's philosophy emerge from a Socratic beginning, as dazzling as a butterfly from a cocoon. We see precursors of his later ideas. Plato is becoming so fascinating that we are likely to forget poor Socrates in our admiration. Let us not forget that the exchange between Meno and Socrates is directed at Socrates. How did the argument affect the original Socratic *elenchus*?

We note that *Meno's difficulty was designed to undermine the Socratic form of skepticism*. It is conceivable that this difficulty was developed contra Socrates in the sophistical School of the Sophist Gorgias, set up in Larissa, to which Meno belonged. What would Socrates have said in his earlier voice?

Plato, who admired Socrates, must have keenly felt the sting in Meno's paradox. Socrates had spent his entire life pursuing knowledge that he hoped would yield wisdom. Plato cannot have Socrates responding to the dilemma proposed by Meno by saying, "you are quite right, Meno, let us give up the search for wisdom, since I cannot know that which I seek, just as Gorgias, the relativist, has claimed." Socrates would have had to face a more serious consequence. He sought knowledge of how to live. To give in to Gorgias now would allow such knowledge to be out of his reach—and out of our reach. Even if there is such a thing as wisdom, we cannot know it (as Gorgias reportedly said.) Each of us can make up our moral wisdom. Plato thus rescued Socrates from Gorgias' imagined critique, but he did so by allowing a new kind of doubt about Socratic ignorance. His resolution of Meno's dilemma may also be expressed as follows:

> Plato: *"Socrates, you have maintained all your life that you seek important knowledge but know nothing. You claim that you are wise in this, that you know nothing. But if you know nothing, how can you know this, that you know nothing? Your remarks may be charming in a self-deprecating way, but are they true? Perhaps you know more than you realize: you may possess wisdom, in a manner of speaking, but not in the manner which allows you to see it easily."*

With such a comment, Plato is not clamping down by an argument "of iron and adamant" that Socrates knew what he claimed not to know. Plato has not shown that Socrates knows wise things. Plato would need to argue much more to show it. He does show that Socrates *may* know more than he realized. That is enough. Socrates claimed not to know. He thought he was wiser than others because he at least knew that he did not know. But now it turns out that he did not know that even that! Later, this became a slogan for skeptical Platonism in the later Academy.

Meno's dilemma may be avoided if there is a distinction between two forms of knowing, i.e., knowledge by possession and knowledge by recollection.[2] His avowed ignorance is compatible with his knowledge of the answer he seeks, provided that he was unaware of his own wisdom. He always knew in one sense but did not know in another. It follows that Socratic skepticism involves a false claim. Socrates thought that he could allow that he was the wisest of all, because though he knew nothing, at least he knew that he knew nothing, which others did not. But perhaps he was mistaken in this claim. He may know more even than he realized.

Socrates does not know even that he does not know. This is Plato's new form of skepticism, more radical than the skepticism of the Socratic craft. We do not know, and we do not even know that we do not know. Later Academic skeptics use this formulation, I suspect, as their tribute to Plato. It was a piquant way of expressing Plato's advance in philosophy. Plato's critics considered it a form of dogmatic assertion, namely a negative dogmatism. But Plato did not assert that Socrates knew anything. The doctrine of recollection is introduced as a hypothesis. Socrates *may have known* more than he thought, refuting his claim without making a dogmatic assertion. That is how the method of hypothesis is first used in Plato's *Meno*.

Plato scholars will point out that Plato does not offer a critique of Socrates—at least, it is not so stated in the Meno. He is defending his teacher. He does not describe knowledge as that which is unconscious in the soul. To know, we must recollect. Moreover, it is not enough to just recollect some one thing, as the young boy did under Socratic questioning. To know mathematics is to know many such things as the ratio of a side of a square's diagonal to its sides. It is also to maneuver confidently among many such bits of wisdom. Socrates was not a mathematician. He did not know all that was recently discovered about the relation of the diagonal of a square to its sides. But Socrates could find his way around ethics and politics. Whatever understanding of knowledge Plato had developed, it would be difficult to claim that Pythagoras knew his mathematics, but Socrates did not know his ethics. Plato's new Socrates in the Meno is mistaken in claiming that he knows that he knows nothing,

> By showing how Socratic skepticism may be mistaken, Plato invented a new form of it. The new skepticism, however, is embodied in his method of philosophizing, not in new claims made by a character in his dialogue. Plato makes no claims about refuting Socrates in this dialogue or elsewhere in his dialogues. But the *elenchus* changes subtly in and after the *Meno*.

The new Platonic dialectic is not personal or *ad hominem*—it became impersonal in a certain sense. A Socratic dialogue is intensely personal. It is conducted about what is both important to an interlocutor and trusted true. Plato's new method examines abstract *hypotheses*—typically, opinions far removed from what is important and true. It has two parts to it. Using the "method of hypothesis" (as found in mathematics, he wrote), we assume that a hypothesis is well understood and accurate and use it to shed light on things—that is, to shed light on whatever follows from it. The new method of the *elenchus*—Plato's *elenchus*—examines a hypothesis of this kind to decide whether it is true. Socratic ignorance, or the doctrine that Socrates knew he did not know, is important to Socrates, who trusted its truth. Plato examines it impersonally and finds it wanting. In the *Meno*, however, he makes no statements about the truth or falsity of Socratic knowledge of ignorance. He allows the issue to settle itself, or perhaps unsettle itself, during argument.

Plato's new *elenchus* assumes nothing to be true when it is under investigation. Nothing contentious is "given" or known while we inquire into it. The Socratic injunction to examine oneself takes precedence over every claim to knowledge. But it is no longer restricted to important matters of personal knowledge. It extends to abstract hypotheses, even general opinions. Plato begins to study issues far removed from the intensely personal knowledge of interlocutors, or even those of Plato. Take the sum total of all that Plato wrote for posterity and ask, "What did Plato really believe (or thought he knew) as a philosopher?" We would be hard-pressed to say exactly. None of his views about first principles are in his voice. He describes them in the voice of a character in a play. Plato continued to use the dialogue form in his explorations of philosophy. He remained a skeptic—a fallibilist. These are not words he uses to describe himself, but they are good words for us to describe him.

For instance, the Forms, or Ideas, are introduced in the *Phaedo* and well represented in the *Republic*. Ideas dominate his dialogues for a stretch of his life. Plato's classic statement of a certain kind of realism and his description of knowledge and ignorance are widely known. The *Form* of a human exists apart from humans, whose similarity (as being human) is to be understood in terms of their participation in the Ideal Human. The separation of a timeless Form and its time-bound participant is a theme in his writing. It began when he wrote the *Phaedo* and continued until he questioned it later in life. When he questioned it later, did he do so merely as part of his Socratic duty to examine himself? Did he entertain doubts? Did he abandon or modify some of his opinions thereafter? It is difficult to know.

In his *Parmenides*, we find a sustained critique of Forms. Plato even introduces the "Third Man" argument. If we need an ideal human, or the Human Form, as a distinct reality to explain the similarity between human beings, we need yet another Human Form to explain the similarity between the Ideal Human Form and time bound humans. The demand for Forms leads to a vicious infinite regress. This argument was later admired by Aristotle, who adopted it. He renounced the separation of Platonic Forms from the things of which they are forms. But this argument results from Plato's (Socratic) active fallibilism: the better his knowledge, the greater his effort to examine it critically. But we cannot reliably determine whether he continued to believe in the Forms as they were portrayed before the *Parmenides* was written.

In the *Sophist*, written after *Parmenides*, he also proposes a criterion for physical existence with startling implications: a thing is real only if it can affect or be affected by another object. Of course, only temporal things can push back when you push them. Forms and Ideas are timeless, so they cannot push or push back. Plato considers whether this is a reason to renounce our knowledge of Ideas or Forms. He thinks of a response to this difficulty, but what Plato endorses is unclear. If Forms are abstract, i.e., changeless and atemporal, how can they affect or be affected by temporal entities, i.e., us? How can we know or grasp them if they cannot change? Would the internally perceived Form not differ from a previously unperceived Form in one respect, namely that it is now perceived where it was not before? Like many other riddles that Plato considers and discusses, his difficulty with Forms has remained with us. If Plato was not convinced by his arguments, many later philosophers were.

We do not know if Plato gave up his earlier account of Forms and if and how he modified them for himself. Some texts written after the Parmenides suggest that Plato was still working with an older theory of Forms. We are left uncertain about Plato's intellectual commitments or his personal belief. Surely, we think, he believed in some version of the Forms at some point in his life. He would not have written all his life of them without some such personal belief. There is ample evidence in Aristotle's writing that he understood Plato to have subscribed to a "Platonic" account of Forms or Ideas. There is also considerable corroboration of this in other commentary on Plato. But which version of his critical thought did he endorse? We do not know. However, he never leaves us in doubt that he remained committed to inquiry about such matters. In this matter, he is relentless. He was a fallibilist, of course. No other fallibilist pursued locating his errors with so much zeal. We may therefore take Plato to be primarily an inquirer or, literally, a skeptic, a zetetic.

Plato did not face one difficulty he should have done if he had known of it. There is an unresolved paradox in the active fallibilism that he practiced. If we challenge a trusted knowledge claim, do we claim it as knowledge when we challenge it? If we trust it as knowledge, we must proclaim its truth. But there is no point in questioning it if we do not consider its falsity. At that moment, are we claiming its truth or allowing it to be untrue? When we turn to skepticism, or the claim that we do not know, is that claim held to be true when we are examining it? There is an air of paradox about a skepticism that is denying itself. Given his philosophical temperament, it may be surprising that Plato did not discuss this difficulty. One must conclude that it did not occur to him.[3]

It would not occur to him if he were a situational skeptic. A situational skeptic would doubt in one situation (let us say in an academic seminar) what he would apply as knowledge in a (practical) case. He did not renounce his beliefs or stop acting on them merely because he was willing to be self-critical. We note that he went to Syracuse in Sicily to advise his friend Dion on governing the city. Plato went there after he wrote in the *Republic*, whose doctrines were subject to severe critical scrutiny in the *Parmenides*. He endorsed doctrines in Syracuse that he was willing to question later on in his "Academy".

IV

Plato taught outside Athens in an orchard known as his Academy. The Academic philosophy taught in his school after him was later seen to be like skepticism by other self-proclaimed skeptics, the Pyrrhonians. In many respects, the arguments of Arcesilaus, an Academic, and Sextus Empricus, a self-proclaimed Pyrrhonian skeptic, are very similar. Presumably, the later Academics, beginning with Arcesilaus and the ("Second") Academy, taught Plato's main teachings as a self-critical fallibilism, a kind of skepticism. They must have taught it as a way of inquiry or a method for students to learn (what we call "critical thinking.") The skeptical reading of Plato's philosophy is also very ancient.[4]

We may separate schools or institutes of learning into two kinds. There are those that someone establishes to teach a doctrine—a pearl of wisdom in Hellenic times or a religion in medieval times. Denominations may be called, and were called in Hellenic times, "schools of thought." Cynics, Cyrenaics, Stoics, and Epicureans each formed such Hellenistic schools. Some schools or institutes of learning do not limit themselves to a body of doctrine or teach a way of life. They try to remain outside the fray. Modern schools are primarily of the latter kind. Most distinguished Colleges, Universities, Institutes, or Academies that we find today foster advanced learning without any alignment with partisanship. During late antiquity, such neutrality was unusual.

In any preeminent academic institution today, each department is devoted to a subject of study. It will collectively constitute a group of thinkers like the philosophers in Plato's Academy, though often more limited in their study. What they have in common is that they will claim to be uncommitted to any school of thought in their field. For this reason, we are surprised to find a modern form of institution under attack for being philosophically partisan, although it does happen. Even religious Colleges formed to teach a prescribed point of view—a faith—are more open to allowing academic freedom to their colleagues than older schools of thought. However, in Plato's day, his school was uniquely nonpartisan. In this respect, it was not a school of thought.

Since it did not endorse a way of life, it came to be challenged as illegitimate by the other schools. It was impractical. "Apraxia" is a charge associated with the Academy not because the members could not look after their practical affairs but because they had no school wisdom to propound. Other schools interpreted Socrates' primary teaching as an endorsement of a form of learning, whereas the Academy taught the method of employing the *elenchus*. Academic skepticism led away from endorsing any wisdom. It showed how any school or institute can avoid becoming a school of thought. It may be called "the skeptic way." A self-proclaimed school of skeptics later officially endorsed a similar method as their form of wisdom. However, the Academy of Plato's successors was not a school of thought. The Academy did not identify itself in this manner. It did not proclaim itself to be a school of skeptics.

Plato's doctrines, whatever we think they were, and whatever principles he discussed in his dialogues, are much like the topics often discussed in academic seminars today. Seminar discussions are covered by a skeptical umbrella, even if people attending are committed to opinions in matters under discussion. When under discussion, members who assert opinions in dispute will withhold assent until the seminar ends. The same people who first declare their endorsement of a statement may raise objections to it and thus test it. That is how any discussion can work to fulfill its academic function. Plato's dialogues are early examples of this self-critical technique.

Taking part in academic discussions is to be a skeptic in this situational way. It follows that the philosophy known as the Academic Philosophy in Hellenistic Greece was, by its very intention, not a school of thought but a self-examination of the kind recommended by Socrates.

V

Francis Bacon's thought seems to have been shaped in very particular ways by the Academic philosophy. From Arcesilaus to Philo, or to the later Academies, there was an attempt to defend an Academy that was not a school of thought but a respectable institute for impersonal teaching. It taught critical thinking, or how to deploy the *elenchus*. It was not an easy practice to defend. It had to avoid the charge of *apraxia*. The followers of Plato's active fallibilism were impractical because they did not teach anyone how to live. They could not, because their teaching was to show students how to be critical of any such claim. The Academics separated examining a philosophy, which was the primary teaching of their academy, from endorsing a practical way of life based on any particular doctrine. Other schools could teach their students how to live. The Academy taught them only how to think critically. They could refute what the other schools taught. They could even examine what they had taught themselves. But they could not teach how to live wisely. They did not know.

It seems they could not have become more practical—not if they were being logically coherent. Either they knew the wisdom they expound or not. If not, then they had no knowledge of which kind of wisdom to impart. Their teaching is impractical, as Stoics charged, if they did know some wisdom to impart. But they knew it, they could not also take it up for criticism, as if they did not know. Thus, Plato's Academy could not be defended as a proper way to teach philosophy in its own time. This failure must have been felt acutely by Academics.

The Academy from Arcesilaus to Philo of Larissa taught critical thinking, as we call it today. Several Academics were also fallibilists, which trait Philo of Larissa later claimed to be common to Plato's philosophy and the Academics who followed. Many Academics are also known to have endorsed ethical and political doctrines. Plato, for instance, traveled to Syracuse in Sicily to advise his friend, its leader, on how to apply the doctrines that he had proposed in the *Republic*. But he also wrote a dialogue, the *Theaetetus*, examining its assumptions about knowledge. Cicero, who adopted Academic fallibilism, following Philo of Larissa, also wrote a book endorsing Stoic ethics, though he did not endorse its epistemology. He was not a philosopher primarily but a rhetorician, so we should not expect to find his statement reconciling the Stoic Sage with the Socratic *elenchus*. Combining an ethical doctrine of any classic school with fallibilism requires careful argument and mutual accommodation because without it, they are together logically inconsistent.

The Stoics had proposed a new philosophy to rival that of Aristotle and Plato in some important details. In Stoicism, there was announced a new conception of ethics, one based more strictly on deterministic virtue, somewhat as in Socratic philosophy, more than in Aristotle's or Plato's writings; a conception of nature as a plenum, and an epistemology that claimed that wisdom of the kind that Socrates had hoped to obtain could be learned from apprehensive or kataleptic perceptions and by living in harmony with nature. In these matters, Zeno the Stoic differed from Plato, the skeptic.

When we perceive a thing, it forms an "impression" or a "stamp" on us, just as a seal makes an impression when stamped on a block of wax. Plato describes this in a famous passage in the *Theaetetus*. Among such impressions, some are from an object we perceive so well that we "grasp" the thing or fully apprehend it by the mind. These are called "kataleptic" impressions. "Katalepsis" or the idea of "grasping" may be suggested by the notion of knowledge (like a bird) in one's hand in Plato's image of the bird in the aviary. When we grasp it, we know. Unlike Plato, the Stoics claimed that katalepsis is perceptual (Long and Sedley 1987, 236–59).

The notion of either a perceptual or conceptual grasp or apprehension is vital to understanding the new scientific method. Francis Bacon calls it "taking hold of the thing." (Bacon, I.13). It is also a tactile metaphor for visual perceptions and mental operations. It indicates that we "get a hold of" the thing by a mental act. We grasp what it is *in reality* rather than merely *touching upon it* superficially by verbally referencing it. Such an apprehension or proper knowledge is the basis of wisdom, according to the Stoics. The Sage (a Stoic ideal of the wise person) relies only on kataleptic impressions to form opinions, whereas we lesser mortals may form opinions based on impressions that are not kataleptic, which is why the Sage is Sage and we foolish. Sextus Empiricus later suggested (as the Academic skeptics had) that since appearances can deceive, the Sage cannot remain a Sage for long.

Cicero imagines how Arcesilaus must have criticized Zeno the Stoic on this doctrine, just as Socrates must have criticized his interlocutors. The gist of this criticism is that even if there were such a kataleptic impression, we could not know it to be so. "I think I grasp that the person walking towards me is Socrates," I say, but do I grasp that it is Socrates? It may be his twin, indistinguishable from Socrates. If I mistake the impression of Socrates to be kataleptic, I would be committing an error. No criterion within the appearance marks it as true and incapable of error (Cicero 2006/45 BCE, 45 or 2.77).

Two things are important to note: first, the notion of katalepsis that was introduced by the Stoics and challenged by the skeptics found a particular solution in the work of Francis Bacon; second, the critique of katalepsis, which is a part of the Academic attempt to defend situational skepticism, or fallibilism, becomes an important part of Francis Bacon's method of science. There is also that third thing of importance, of course, which is the paradox at the heart of Academic skepticism, but it was not addressed within the Academy in a general way. It remains an unresolved difficulty. They tried to finesse their favorite ethical theories as "reasonable," "probable," "verisimilar," and so on—but fine distinctions between these phrases do not solve the fundamental problem. If we argue that we cannot know the truth of any principle, then we cannot know its probability by the same argument. The trouble runs deep.

The paradox of doubting our abstract knowledge while endorsing it—i.e., the paradox of fallibilism—remained unresolved. Plato avoided the problem by practicing situational skepticism. If we separate practice from theory, we can prevent the problem from popping up, but only up to a point. Sooner or later, the issue pops up when principles are put into practice. In any case, if we formulate a skeptical theory of knowledge, we need to face up to it. Plato had not developed his form of skepticism or fallibilism as a doctrine explicitly for further study, so he did not seem to have seen the difficulty. At the end of the scholarch Philo's career, he fell out with Antiochus on how skeptical or dogmatic the Academy should be. As Cicero tells us, the schism was profound. There is no point in deciding which of them is right because they fell out on a fault line that neither could erase (Tarrant 1985).

The Pyrrhonian skeptics were a splinter group from the Academy who recognized the difficulty in all its generality. We must turn to their writing, particularly the writing of Sextus Empiricus. Being self-proclaimed skeptics, Pyrrhonians could identify the problem. They found a solution. However unsatisfactory we may find it, their answer was ingenious. They gave up ever *claiming* knowledge of any hidden reality—but they gave up their claims without admitting that real knowledge cannot be attained. They did not claim that either. To say that we cannot know anything would be paradoxical. How could we know that nothing is known? Their cunning solution was to recommend an endless search that could not succeed. Their cunning solution had a cost. They had to give up the search for knowledge as the skeptic's goal, which had motivated Socrates and Plato and still interests us.

Notes

1. Peter Harrison makes this point in his (2023) and other writing. The shape, however, may be changed today without violating any presuppositions of modern science.
2. Plato does not make this distinction until the *Theaetetus*, where he discusses the mind as possessing Ideas (and Eidola) like birds, as in an aviary.
3. John Post (1970) has shown that Plato's form of skepticism is paradoxical. The paradoxical claim was often repeated in the later Academy: "We know nothing, not even this, that we do know nothing." This turns out to support a semantic antinomy. John Post's logical result has interesting consequences that have not yet been fully explored.
4. Harold Tarrant (1985) studied the skeptical and the affirmative interpretation of Plato's doctrines in the later Academy, when dogma overcame skepticisim as Plato's true legacy. But Plato as an Academic is a recurring theme in the history of philosophy. More recent, for instance, is Katja Maria Vogt's (2012) erudite interpretation.

References

Cicero. 2006/45BCE. *On Academic Scepticism*. Tr. Charles Brittain. Indianapolis: Hackett.

Harrison, Peter. 2023. Normativity and the Critical Functions of Genealogy: The Case of Modern Science. *Modern Theology*. https://doi.org/10.1111/moth.12857.

Long, A.A., and D.N. Sedley. 1987. *The Hellenistic Philosophers, Vol. I: Translations of the Texts*. Cambridge: Cambridge University Press.

Post, John. 1970. The Possible Liar. *Noûs*, v.4 # 4.

Sprat, Thomas. 1667. *The History of the Royal Society of London, for the Improving of Natural Knowledge*. London: T.R.

Tarrant, Harold. 1985. *Scepticism or Platonism?* Cambridge: Cambridge University Press.

Vlastos, Gregory. 1994. The Socratic Elenchus: Method Is All. In *Socratic Studies*, ed. Myles Burnyeat, 1–38. Cambridge: Cambridge University Press.

Vogt, Katja Maria. 2012. *Belief and Truth: A Skeptic Reading of Plato*. New York, N.Y.: Oxford University Press.

4

Pyrrhonians, a School of Skeptics

I

A new school of self-described "Pyrrhonian" skepticism sprang up around the time of Philo and Antiochus. Pyrrho of Elis, to whom the dissidents deferred, was in the entourage of Alexander of Macedon. When Alexander went east to conquer Persia, he took some philosophers with him. Pyrrho was among them. They went as far as the Indus River valley. Pyrrho's opinions, however, were probably not the origin of the later Pyrrhonian philosophy. Richard Bett has shown many differences between the unwritten opinions of Pyrrho, as reported by his friend and student Timon, and Aenesidemus, the dissident who broke off from the Academy. Our record of their writing is not extensive, as most of it is lost. But the evidence Richard Bett has marshaled seems strong. There are also some differences between the writings of Aenesidemus and those of Sextus Empiricus. Our interest, however, is in the writings of Sextus Empricus, which have been preserved for us. It is simplest to treat his writing on "Pyrrhonian skepticism" as a dissident form of the later Academic philosophy.[1]

J. Hattiangadi, *Francis Bacon's Skeptical Recipes for New Knowledge*,
https://doi.org/10.1007/978-3-031-52585-8_4

The most important dissident trait among Pyrrhonians is their proclamation of skepticism. They self-identified as a school of skeptics. In Plato's Academy, skepticism was only practiced, not proclaimed. It was even taught by example but apparently not isolated for analysis. Aenesidemus set up shop as a dissident. It is not unreasonable to speculate that the Pyrrhonians were shocked by their reading of the dispute between Philo and Antiochus, which had divided Academics. They thought that the Academy had not faced up to their difficulty. The Academics deny knowledge of any principle while endorsing one at the same time. If so, then it is not surprising that they invoked Pyrrho—in other words, they invoked Pyrrho to abandon Plato and his quest for knowledge. They claimed that they sought peace of mind or *ataraxia*. What was the headache, we may ask, that they wished to avoid? It does not seem unreasonable to suppose that the dispute between Philo and Antiochus would have been a prominent candidate.

II

Few books have as arresting or misleading a beginning as Sextus Empiricus' *Outlines of Pyrrhonism*.

"When people search for something, the likely outcome is," he wrote, "that either they find it or, or not finding it, they accept that it cannot be found, or they continue the search. So also in the case of what is what is sought in philosophy…." (Sextus Empricus 1995/second century, 89) He identifies as Dogmatists those who stop because they have found what they seek; the Academics conclude that it cannot be found and therefore give up; the Skeptics continue to seek. Academic Philosophy comes off as a negative dogmatism, while the Skeptic philosopher continues to inquire.

This promising beginning entices us to seek answers to inquiries. Sextus Empiricus suggests that of the three, only Pyrrhonian skeptics allow inquiry into knowledge. But this description is misleading. Sextus Empiricus does not inquire into furthering knowledge of any kind—not in the book that follows this introduction nor elsewhere in his writings. Little curiosity about anything is to be found in his work. His book is an extensive description of what Pyrrhonian skeptics say and do to defend their skeptic school and what a skeptic could say in response to any claim describing a hidden reality. Knowledge is essentially hidden, though they cannot say so. The descriptive task of what the skeptics say and do is conducted in the third person with great attention to detail. It is admirably clear. Despite the opening paragraph, he promises nothing of that spirit of inquiry. He kills inquiry—but in the name of continuing inquiry. The point of Pyrrhonian investigations is not to find knowledge of hidden realities. It is to keep us from admitting that we cannot know what we seek. We know that but cannot say it! Carneades, the Academic, said nothing can be known, which is paradoxical. (Does he know that?) The Pyrrhonians do not say so, for they have forsaken negative dogmatism and keep inquiring.

Sextus Empiricus depicts a school of thought called "Pyrrhonian skepticism." He is defending it as thoroughly as rival schools of thought do, i.e., the dogmatic schools. Forming a school is another advance over the Academy. He is doing what is required of a professional skeptic or a skeptic of the schools. His outline is an outline of the doctrinal rule of the skeptic school:

> We … [say] in replying to the question "Has the skeptic a doctrinal rule?" "…[I]f one defines a 'doctrinal rule' as adherence to a number of dogmas which are dependent on one another, and on appearances, and defines 'dogma' as 'assent to a non-evident proposition' then we shall say that he does not have a doctrinal rule. But if one defines 'doctrinal rule' as 'procedure which, in accordance with appearance, follows a certain line of reasoning, that reasoning indicating how it is possible to live rightly… and tending to enable one to suspend one's judgment,' then we say that he has a doctrinal rule. For we follow a line of reasoning which, in accordance with the appearances, points us to a life comfortable with the customs of our country and its laws and institutions, and to our own passive instinctive feelings." (Sextus Empiricus 1990/second century, 20–21)

The main difference between the skeptic of the Academy and the Pyrrhonian skeptic is the following: The skeptic of the Academy would assent to a proposition if it is reasonable, as Arcesilaus said, or plausible, as the famous Carneades allegedly described it. Yet, they would not assert its truth when it is under investigation. Situational skepticism, of course, allows an examination of our most cherished beliefs in a philosophical inquiry. Not so the Pyrrhonian skeptic, who will assent to whatever is "apparently" as true, but he would deny it as true of non-evident reality.

Thus, if we said, "The sun is really a nearby fixed star," both the Academic and the skeptic would say that they do not endorse this. However, some Academics might agree that it is *plausible to suppose* that the sun is a nearby fixed star if it makes sense of the planets, though they would hasten to add that it is not known. They would withhold assent but may allow that it a reasonably good astronomy. The Pyrrhonian skeptics, however, would say that *it would not seem or appear to be the case* that the sun is a fixed star. Assent must be withheld since the claim is merely an opinion, and other opinions may also be offered. Sextus Empiricus takes great pains to show the difference between his and the other skeptical claims, suggesting that only his own is genuinely skeptical.

Sextus Empiricus tells us, all the same, that his views are much like Arcesilaus' in many respects, even though Arcesilaus was the first Academic (unless it was Plato himself.) Pyrrhonians and Academics differed in distinguishing opinions into those asserted and those that were not. This difference seems subtle but worthy of note. Academics asserted opinions that they were willing to examine in the right situation. Sextus Empiricus suggests that the skeptic asserts what *appears to be true* but nothing about what *is true in reality*.

The Pyrrhonian skeptic way is as follows. Some may oppose things as they appear to us with the same things according to their hidden reality. So, for instance, that tree seems to me to be large. I may say that a tree is large because it has many atoms, though the atoms are not evident to us. A skeptic consents to the first claim about the tree and declines the second part of the statement about the non-evident atoms. When anyone claims to know hidden realities (if they are being "dogmatic,") the skeptic has a method to help us overcome dogmatism. First, we oppose one dogma with another dogma. Due to the balance between opposed opinions (equipollence, *isosthenia*), the skeptic can suspend judgment (called *epoché*.) Deciding to withhold assent from making judgments removes the vexing need to decide on reality. Abstaining may help us achieve peace of mind or *ataraxia*. That is the skeptic's way, as described by a skeptic, Sextus Empricus.

Sextus Empiricus concedes that Arcesilaus, too, held a similar theory of the way forward. Arcesilaus, scholarch of the Second Academy after Plato, had also suggested that we should balance claims about reality against one another. Such a balance led him also to *epoché*, which enabled him to withhold his assent. The main difference, therefore, lies in the last step. For the Pyrrhonian skeptic, the goal of inquiry is *ataraxia* or peace of mind. Arcesilaus said nothing of this. His contemporary, Pyrrho, had asserted *ataraxia* as his goal. An Academic skeptic's aim, like the Socratic goal, is always to gain knowledge of wisdom. Nothing less is worthy of philosophical pursuit.

Sextus Empiricus portrays Academics as negative dogmatists. He picks out Carneades and Clitomachus, who came after Arcesilaus in the Academy.[2] Carneades did not set down his thoughts in writing. In his day, he was famous for arguing for one opinion persuasively and then for its contrary just as persuasively. He visited Rome and created a sensation by doing so in the Senate. Persuasion may not have been his goal since he would have refrained from endorsing either persuasive opinion.

Clitomachus, reporting Carneades' thoughts, wrote several books (now lost) that Cicero had read and admired. Cicero was much impressed by the arguments in them. Titles of two of Clitomachus' books are *On the Withholding of Assent* and another called *On the Schools of Thought.* It would seem from these two titles that he was, like Sextus Empiricus, a critic of dogmatism. The title of one book also shows that he withheld assent rather than claimed that he did not know. The second book's title indicates that he did not consider Academics a school of thought. He could have said, "Other schools of thought." We see that Sextus Emprivicus endorses the withholding of assent, but he wants his way of thinking to be a school of thought, too—though not a school of dogmatic thought. The Pyrrhonians lay the charge of negative dogmatism that the Academics probably did not earn. They probably did not formulate their skeptical stance at all, or not explicitly. The charge arises, very likely, out of a misunderstanding of a turn of phrase.

Sextus Empiricus seems to misunderstand a claim that was likely routinely made by the Academics. Arcesilaus claimed that he knew "nothing, not even that he knew nothing." Carneades is reputed to have declared "that nothing can be known, not even this, that nothing can be known."

Sextus Empiricus fails to note that these expressions are merely slogans about what they learned from Plato, who went even beyond Socratic ignorance in the *Meno*. They were not intended as claims about the nature of skeptical knowledge. Academics were not negative dogamtists, because they did not insist on the truth of their famous dictum. There is no evidence that they refused to critically evaluate it. But the dictum was a symptom of a deeper problem of skepticism, which the Pyrrhonians recognized and learned to avoid. That recognition is their single greatest contribution to skeptical inquiry.

The lesson of the *Meno* may be expressed thus: We need to be even more humble than the humble Socrates, who said with so much humility that he had no knowledge and was wise only because at least he knew that he was ignorant. But he did not know even that he did not know. We may know in one sense of the word (i.e., innately) but not in another (as recollected.) But (and Sextus Empiricus misses this) this is so *because we (or Socrates) may know the thing we seek, contrary to what we (or Socrates) may think*. Sextus Empiricus misses that thought in the *Meno*, to which the Academics must have been alluding. Since the *Meno* reflects Plato's philosophy, when he went beyond Socrates, it cannot be a negative dogmatism. Plato was a situational skeptic, a philosopher who unhappily failed to see the paradox in his practice because he did not formulate it thus.

The Pyrrhonians were right to criticize Academics for not providing a philosophical solution to the question of fallible knowledge. Those who first assert something and then deny it contradict themselves. Their incoherence needs resolution. If they restrict themselves to appearances—and only as appearances—then there is no incoherence. Something may seem to be so and yet seem not to be so. Both are possible if the circumstances differ. What seems little from a great distance may seem large when close by. Nothing is large and small, as it can appear from different vantage points. Appearance is very forgiving. The Pyrrhonian skeptics found a way of avoiding self-contradiction for skeptics. But it came at a cost. They no longer sought knowledge because it was paradoxical, though they would not say that seeking knowledge must stop, for that would be negative dogmatism. Negative dogmatism would contradict skepticism, which is their way of inquiry. They claimed to never give up on inquiry to arrive at their deceptive form of skepticism.

The idea of a skeptical school of thought is a degenerate or corrupt form of Academic philosophy. It violates Plato's insight into the academic enterprise that we seek knowledge but recognize our fallibility. Nonetheless, Sextus Empiricus commands our attention. By its very degeneracy, this corrupt form of Plato's philosophy gives us a better insight into Plato's skepticism. Pyrrhonians did better at what the Academic philosophers said for themselves because they were at least logically self-consistent.

The reason why they called themselves "Pyrrhonian" seems to be this. Pyrrho traveled with Alexander of Macedon during his triumphant journey of conquests across Egypt and Persia to India, and there near the Indus river, he came across the "gymnosophists" or the naked philosophers—very likely the Digamber Jains, i.e. the "sky-clad" Jains. It could also be a skeptical brotherhood of Buddhists, as commonly supposed, but nakedness was not one of their communal practices. Pyrrho had just come from Athens, where one Socrates, a skeptic, had shaken up the wise men of their city. But here, Pyrrho found a whole community of naked skeptics.

Returning to Athens, Pyrrho was a confirmed skeptic. He had adopted what may have been a Jain goal of achieving "Moksha," freedom from worry or *ataraxia*. This goal shared with Epicureans, not unlike the Stoics' *apatheia*, is also related to tranquility or peace of mind, differently construed in each case but showing an Indian influence. The Jain ideal of "Moksha," a state reached by a person free of suffering, is also paralleled in late Brahmanical and early Buddhist thought. In Pali, it was called "Nibbana." A permanent state of relief from suffering as their goal may have been the most widely felt and lasting influence of Alexander's conquests on Hellenistic philosophies. Hellenistic Buddhism also had a lasting influence on philosophy in what is now Afghanistan and Pakistan, in the Gandhara tradition.

III

Pyrrhonians produced no later skeptical philosophers that are remembered today. Surprisingly, it was a school mostly among medical practitioners. Medical practice was one of the first sciences to specialize and sever from general philosophy. After Hippocrates, it remained a philosophically sophisticated practice. Medicine acquired many specialized research areas. More recent sciences, such as physics, chemistry, and biology, followed a similar path after the eighteenth century. But the philosophical component remains with each. Sextus Empiricus himself was a medical practitioner. Like Galen or Soranus, his interest remained keenly philosophical. His brief remarks on the topic of medical practice are revealing about Pyrrhonian skepticism.[3]

In ancient Greece, two prominent schools of medicine survived through Roman times. One of these is the school of "Physiologoi," which studied the true natures of things, such as the hidden causes of disease (etiology) (Galen 1985/second century). The others were Empirics, who did not. To understand the difference between the two, consider the diagnosis of an ailment in terms of a fundamental condition based on overt symptoms. A "scientist," to use a modern word, will attempt to tell us what is wrong and what needs to be prescribed to the patient for a cure. On the contrary, the empiricist will say we can only know the symptoms or appearances. Therefore, the non-evident causes adduced by the physiologoi (scientists/philosophers) must be studiously disregarded in finding the proper treatment. Non-evident causes are not known. Pyrrhonians are consequently unhappy with the philosophical, or as we would say today, scientific study of medicine.

There are two kinds of treatments notable in medicine. One provides "symptomatic relief" when available. The other provides a "cure." A cure changes the underlying condition of the ailment and thus also relieves the symptoms. The empirics did not believe in real cures—because only symptomatic treatments can be known. If a syndrome has been identified and a well-known remedy is indicated, the empiric prescribes it. The *physiologoi* are different. They would first investigate the underlying condition or cause (*etiology*) and then specify a remedy that would cure. They would prescribe symptomatic relief if they cannot find the cure.

The third school of medical thought was of later and Roman origin, called the Methodic school or Methodists, based on its indigenous tradition of medical treatment. The "method" of the Methodist was two-fold: they relied on the study of the disease itself (pathology). They also relied on tests or techniques to identify a medical condition and the right remedy. Each pathology develops, with appropriate treatment methods at every stage. Such methods or techniques were emphasized in the school.

There were also methods to test the effectiveness of a prescribed treatment. So, for instance, there was a method to test whether a wet nurse is suitable to breastfeed a neonate. A drop of her milk was placed on a fingernail. Its color and apparent viscosity on the nail would determine whether the milk was suitable for feeding the neonate. Such tests or methods were their lasting contribution to medicine. However, they also had to classify ailments. They subscribed to a theory of underlying conditions of contraction and dilation of bodies, the state of their pores and how atoms moved among them.[4] There are philosophical ("physiological" in ancient terminology) components of methodism in medicine.

We can identify the three schools of Graeco-Roman medical thought by their attitudes to three parts of ancient medical science. There is etiology, the hidden causes of a disease; pathogenesis, or the development of disease in time; and symptomatology or the clinical manifestation of the disease in syndromes. The Empirics only looked at the symptoms, excluding the other two. The Methodical school made careful use of pathogensis and tests.[5] Only the philosophical (scientific) school of *physiologoi* also allowed a place for etiology or a full causal analysis in its diagnosis and prescription.

Sextus Empiricus claimed that the medical empiric does not think we can obtain knowledge of underlying conditions or non-evident reality. He is giving up looking for a deeper diagnosis. For this reason, Pyrrhonians seem best suited to the empiric school of medicine. In a curious passage, however, Sextus Empiricus suggests that the Pyrrhonian skeptic should not subscribe to medical empiricism (though, in practice, Sextus Empiricus may have been one). He indicates that he would support the Methodic school, but why he suggested his support is unclear.

A curious feature of Sextus Empiricus' support of skepticism is that he finds that if we oppose one medical dogma with another, we will find that we can withhold assent and achieve *ataraxia*. The medical practitioner here achieves *ataraxia*. But he finds tranquility by refraining from the intellectual stress of studying the true underlying cause of an ailment. Would a patient not wish for a better diagnosis at the cost of a bit less peace of mind for the medical practitioner? A more robust investigative approach and a better remedy may provide a patient with more *ataraxia*.

Pyrrhonian scholars did not extend their analysis that far. Perhaps modern science has spoiled us after the late nineteenth century (Bernard 1957/1864). Finding true but hidden causal conditions was not as commonly expected in Hellenistic times.

IV

In his systematic account, Sextus Empiricus offers arguments that assist skeptics in questioning and undermining the opposition. The treatment of the arguments of opposing dogmatists is extensive and carefully written. So good is it that Sextus Empiricus' sketches of past thought are among the most informative reports about several ancient philosophical schools and their typical arguments. His doxographic treasure is equally valuable in preserving the arguments of earlier philosophers as it is in demolishing their conclusions.

Pyrrhonian arguments are marshaled and presented as if they are prescriptive therapy. Their goal is *ataraxia*. Peace of mind, of course, can be achieved by living according to the dogma of epicureanism (as epicureans claimed). However, Pyrrhonians countered that it could be achieved just as well by suspending judgment or *epoché*. To be in suspension, we must free ourselves from controversial opinions. We need to take certain arguments as prescribed. We take them as we may take an aspirin with bed rest. These are the arguments that are presented as a cure for the opposite of *ataraxia*, which is a kind of philosophical turbulence of the mind that a controversy about reality may bring to us. The skeptical Academy fell into a dispute between Philo and Antiochus. If that headache led to the Pyrrhonians seeking ataraxia, it might be an overreaction.

There is a doubt, of course, whether the skeptical arguments deployed against the opinions of others bring *epoché*, which in turn supposedly brings us peace of mind. These are suppositions of Pyrrhonians. Both halves are questionable. There are counter instances to the second. Some people who lose their loved ones find solace in the opinion that the loved one still exists somewhere, just as last remembered. Their supposed continuing existence is nonevident. The bereaved can be challenged by proposing alternative accounts of life or death after death. Contrast other opinions, get *epoché*, you may say, and the bereaved may achieve *ataraxia*. Factual evidence from studying the recently bereaved would suggest that this would not console the grieving partner. *Epoché* can be more disturbing than the possibly controversial opinion to which they cling.

Although bereavement exhibits this fact more saliently, a case may be made that clinging to a dogma may give some people a permanent state of *ataraxia*, also called blissful ignorance. For instance, a religion may be challenged by opposing another to it. But properly tutored, faith-based opinion can be configured to avoid any alternative. The faithful know how to counter it by remaining loyal to their community of faith. Blissful ignorance of this kind will give some true believers peace of mind that they cannot get from entertaining doubts. Such a ploy was discovered by Augustine, Bishop of Hippo. He wrote as a convert to a religion on how to avoid troubling skepticism. Sextus Empiricus assumes that *ataraxia* may follow upon *epoché*. It may. But the faithful may find another way to a promise of salvation, whose contemplation will also bring them *ataraxia*. The Pyrrhonian truth is far from apparent to the convert who doubts the method of doubt based on unquestioning loyalty to a faith.

Pyrrhonian skepticism is captured in two main sets of arguments, though these are not the only arguments explored as modes in Sextus Empiricus's texts. These two arguments are often translated as two sets of "modes" or "tropes." These are sets of arguments named after Pyrrhonian skeptics who collected them from the practices of earlier skeptics and philosophers. They elaborated and refined them into the form they are given in Sextus Empiricus' text. These are the ten arguments of Aenesidemus and the five of Agrippa, respectively. In recent theories of knowledge, the five arguments collected by Agrippa have received more attention—perhaps more than their fair share. However, the ten modes or arguments of Aenisedemus were the more influential on the development of the scientific method in the seventeenth century.

The ten arguments of Aenesidemus are about sensory perception as the medium through which we gain our knowledge of reality. Perceptual knowledge was of great importance. The Epicureans, the Stoics, and the Aristotelians argued that our understanding of reality, which they believed we could obtain, must ultimately derive from perceiving it. They offered different accounts of perception based on their various theories of the physical world and human organs of sense. However, their epistemologies were closer to one another than other parts of their philosophy. The ten modes or arguments of Aenisedemus can be deployed against any of these schools.

Sextus Empiricus divides the ten modes into three kinds, basing those arguments on the subject making the judgment, on features of the object judged, and those that are mixed, or both. Therefore, the arguments are of two kinds, which cover the three modes. Of one kind are arguments about subjective errors. All of them relate to the "relativity" of our opinion (one especially is called the "relativity" argument). Of the other kind are arguments involving difficulties of the things themselves. These may be called "discrepancies" in things. All the arguments of Aenesidemus are one or another or a mixture. Among the relativity arguments, we find the first five of the following (Sextus Empiricus 1990/second century, 27–63 or Book I, section XIV):

1. Differences among perceptual organs and abilities make knowledge of what is perceived true relative to the animal's perceptual apparatus since they would entertain different images or sensory qualities (*Phantasia.*) Human beings will be one among the animals, of course, that perceive reality in their way.
2. Individual human beings will manifest differences in their ability to perceive and acuity and perception, so perceptual knowledge is also relative to personal differences between us.
3. Even for a single person, information perceived from different sense organs will often contradict what another organ perceives—for instance, someone may see a painting as a scene in depth but will feel the same painting to be smooth (and lacking in depth.)
4. The same thing also shows different features in our perceptions of them, over time and in its differing conditions, so one is hard-pressed to say which of these perceptions tells us what it "really" is.
5. Our spatial perspective on things (angle, distance, etc.), media features, and other factors can influence what we perceive as real.
6. We can perceive objects only indirectly through the medium of air, in the case of sound, for instance, and mainly through the membranes of the ear. When we perceive something, it may not be the

thing itself so much as an admixture of the item with the physical medium. Similar is sight and the eye (e.g., a jaundiced eye sees things yellow, which to others seem white, and we may then ask: which is it, yellow or white?).

7. Perceived objects may differ when viewed in slightly varying circumstances—such as when a goat's horn appears white when shaved, but when mashed up horn seems dark. When we scatter sand about, it appears rough but smooth when we gather it into a dune. If we imbibe some things in large quantities, they may be harmful but beneficial when consumed in smaller quantities. All these make it hard to say what such items are in themselves.
8. All things are perceived in relation to other items and we cannot judge them for their reality in themselves.
9. Perceptions of things over many instances will dull their appearance to us, but if we pay attention to a thing when it is rare, it will exhibit features that we may otherwise miss.
10. We often have different opinions based on our upbringing, which influences what we perceive or attend to and, when we do, how it appears to us.

We may find some of these difficulties of perceptual knowing in Plato's *Theatetus*. However, there are also details in the Modes that go well beyond Plato's discussion. They were quite possibly developed by other Academics whose works have been lost, or they may have been a contribution by Aenesidemus.

Their division into the objections from relativity and discrepancy is essential for understanding Bacon's *New Organon*. He rejects all theories of knowledge based on superficial observations because they are prone to relativism. However, the discrepancies in the appearance of things are important to science. They are the hidden source of knowledge that the Pyrrhonian skeptics had missed. They had bypassed their significance because they had adopted too pessimistic a view of obtaining knowledge. Francis Bacon thought that they assumed skepticism prematurely before a more thorough investigation of the facts. Francis Bacon's optimistic method exploits the discrepancies in things to winkle out a more faithful or true schematism (model) of reality.

V

The five arguments of Agrippa are easier to state and understand than the ten of Aenesidemus. The first argument is based on discrepancies, the second on infinite regress, the third on relativity, the fourth on hypothesis, and the last on reciprocity (Sextus Empiricus 1990/second century, 63–67 or Book I, section XV).

The first argument on discrepancies is applied when there is a conflict in how an object presents itself to us. We can fall back upon a suspension of judgment concerning conflicting appearances. For instance, the same thing that appears round from farther away may appear square when judged from nearby. We say that it is round as it seems there and square here. Some arguments from the ten modes of Aenisedemus that Francis Bacon exploited are similar.

The third argument of Agrippa (skipping the second argument until later) is on relativity. It shows how an object's appearance differs for different people or the different senses they rely on. Based on this relativity, we are therefore allowed to suspend judgment on the true nature of what appears to us in these different ways. There is also a summary of some of the ten modes of Aenesidemus. We may note that all the ten modes of Aenesidemus, as Sextus Empiricus himself describes them, may be thus included in the first and third summaries of the modes of Agrippa.

The three other arguments of the five collected by Agrippa (numbered two, four, and five) are more of a piece. They are best understood together and relate to Aristotle's argument in the *Posterior Analytics*. One is from *infinite regress*. In this mode, we show that any claim to demonstrate or prove something requires premises and a valid argument. But the premises need to be true and demonstrated. This demonstration needs another good argument with premises that require further proof, and so on. This sequence of proof leads to a regress to infinity. Infinity is too long to wait. We can never supply the final proof we need to complete the demonstration.

The next mode of argument, *from reciprocity or circularity*, is this: When we set out any proof, we may prove its premises, in turn, by assuming something that was to be proved. This process is circular and undermines the proof. In the simplest case, any statement may be adduced as proof of itself if we allow for circularity.

The last mode of argument *from hypothesis* describes our state when we begin with some hypothesis as a starting point that we do not, in turn, prove. We can start with it as a dogma. We can consider another and contrary hypothesis, which is equally good as a starting point of proof. By seeing that we cannot judge the superiority of one proof over the other, since we cannot assume the truth of an unproved hypothesis, we suspend judgment on both.

The three arguments form a trilemma, showing we cannot demonstrate truth. To demonstrate a truth as known, it must have premises. Premises are either hypotheses or demonstrated. Since hypothetical assumptions are not knowledge, we seek demonstrations of premises. If demonstrations must use new premises each time, we are led to an infinite regress. If we fall back for premises on what has already been demonstrated, we are undermined by circularity or reciprocity.

VI

Pyrrhonian skeptics systematized the separation of opinions into a difference between opinions about what appears to be the case and opinions about what is not evident but is the case. Different members of the Academy may or may not have formulated their thoughts this way by following Arcesilaus' original distinction. It became a permanent and systematic part of the new skeptical school of thought that Sextus Empiricus describes. The skeptic asserts an opinion about what appears to be the case. At the same time, assent is withheld about any nonevident reality. We may put it this in another way: we may rely on apparent knowledge as apparently true without any commitment to anything real as known to be true. Using this criterion, the skeptic can meet the charge that skepticism is impractical by showing how one can "live by appearances." In other words, we can live by relying only on what is apparently true without wondering what is true of non-evident reality.

Appearances are, of course, how reality manifests itself to us. When we live by appearances, in most instances, we are living by reality, too. How a thing appears is evidence of what it is. This relation becomes important to Francis Bacon's thinking about science. Though they may, what seems true and what is true need not diverge. An issue arises only when someone, perhaps a philosopher, claims that though a thing may appear so, its real non-evident nature is different. For instance, an Epicurean may declare that an object, let us say a table, which seems large and smooth, is, in reality, a jagged collection of atoms mutually separated in a void. To any such claim, the Pyrrhonian skeptics respond that they withhold assent concerning that opinion about a hidden reality. Withholding assent is right when we only assent to something known. But they accept the table's appearance. In this way, philosophers' different dogmatic ideas concerning fundamental matters are set aside as of no consequence to the skeptic since all claims about non-evident reality are equally unworthy of assent. But the skeptic can base all practical actions on what the object appears to be, namely a table, disregarding theories of its true nature. A skeptic lives by appearances. So the Pyrrhonian skeptic is very practical—perhaps even too focused on the immediately practical, one may say.

Academic philosophers had once faced a charge of being impractical but in a different way. They did not prescribe a way of life. The lack is not surprising since the academic school was not a school of thought like the others. They lacked commitment to any system of ethical doctrine, though any individual philosopher among the Academics may endorse a way of life. One may even say that they systematically avoid supporting a system of ethics as members of the Academy. They spent their time examining how other schools made their contributions and may even endorse what seemed to be the best among them. But they endorsed it not as knowledge but as laudable opinion.

Some Academics seem to have consented to making some ethical or political claims. However, the ethical basis of an Academic's actions must not be asserted as knowledge, especially in an intellectual or Socratic setting (as an Academic.). Arcesilaus was willing to describe an actionable opinion, a practical maxim, as "reasonable." Carneades were reported using "probable" or "plausible" for such opinions. He meant this presumably as "plausibly true" without claiming to know it to be so. Of course, these opinions can be doubted academically, so we cannot say we know them. Cicero later adopts the criterion of "probability" and "verisimilitude" for actionable opinions used for practical ends. Of course, it began with Plato.

Plato's situational skepticism allowed him to challenge every account that seemed to him to have merit, including his earlier accounts of justice, virtue, and knowledge. But this did not stop Plato from going to Syracuse, in Sicily, to try and influence its politics, per his judgment. He thought that his opinion was actionable or practically efficacious. The Academics were following Plato's example in endorsing dubious doctrines. Situational skepticism and criteria for practical action, like those offered by Arcesilaus, Carneades, and Philo, allow an Academic to take a strong ethical and political stand. For instance, the criterion of "reasonableness" may be used to support a theory of justice that persuades us that Athens is an unjust state, after all, and that its citizens were lacking in virtue in their condemnation of Socrates. Plato, like Socrates, is thus both a moralist and an inquirer at the same time. Knowledge and self-criticism were not identified as mutually exclusive.

By contrast, the skeptic described by Sextus Empiricus is a very conservative individual. A skeptic lives by the local community's customs, laws, and institutions. What we judge to be good must be based on custom. A Pyrrhonian adopts traditional doctrines because there is no way to know better. Such a skeptic cannot claim to know a non-evident moral value. Suppose something is considered virtuous in a given community. In that case, it appears to be so, and a skeptic has no superior knowledge that can do any better than what is assumed in any such custom.

Once skepticism was integrated into a single *school of thought*, it became the basis of its recommended code of conduct. The task of the professional or self-proclaimed skeptic is to find a fit between our way of life and our skeptical doubt. The skeptic fails, and so only the unexamined life can be lived. Sextus Empiricus' skepticism also differs from the ancient Jain skepticism, which Pyrrho imported from India. In *Against the Ethicists*, he asks, "Is there good and evil by nature?" There are two possible answers to this: that there is good and evil by Nature and that there is not. One would think that the Pyrrhonian manner would dictate that we withhold assent and say, as in Pyrrho's dictum, "No more this than that." On this occasion, however, Sextus Empiricus asserts that there is no good or evil in reality. Good and evil are customary in commanding assent, though it remains hypothetical. There seems to be a discrepancy between this stance and his skepticism (Sextus Empiricus 2000b/second century, 117–22).

However, Sextus Empiricus may be trying to avoid a worse discrepancy that would engulf him if he were to assume a good and evil by nature or of a moral reality that we do not know. There are many customs in many different communities. Communities do not always agree on what they assume or proclaim to be good. At most, one assumption can be true. If we allow that some good is natural and exists, then we could be evil, unknown to us. If there is a good, we could be living a life of evil in some respect, even when conforming to custom. This argument is not found in Sextus Empricus' text, but one may suppose that some such argument led him to deny that there is a moral reality. A good physician lives in conformity with the customs of her patients. At any rate, on this question, he did not seek *epoché* and *ataraxia*.

In the interest of pursuing *ataraxia*, Sextus Empiricus does best to claim that good and evil are constructs of local custom. Sextus Empiricus must have known that tyrants who feel threatened will sometimes do unspeakably cruel and ethically indefensible things to the weak and the defenseless. The twentieth century has offered particularly conspicuous cases of this on a large scale. Still, human moral depravity has been observed for a long time, even if the scale of the cruelty was smaller before we had some modern technology in recent times. While local customs may condone some wickedness as inevitable, *ataraxia* seems hard to maintain in its face, even more so in extreme situations.

The Pyrrhonian school is a decadent form of Academic skepticism, even if it was self-consciously consistent in not prescribing a way of life. It was decadent for claiming to carry on inquiry only in the somewhat bizarre sense that no inquiry is prohibited, though certainly none is encouraged that may lead to actual knowledge of reality. Knowledge, says Sextus Empiricus, is possible, and Pyrrhonians have given up the negative dogmatism they say they see in the writings of the Academics. It would seem that even this claim is a bit disingenuous.

There is a test case for the true intent of Pyrrhonism. What does a skeptic do when faced with a theory that the skeptic cannot refute? Does the skeptic allow such knowledge because it cannot now be denied? Here is what Sextus Empiricus has to say in response to someone who puts this new theory forward. "We say to him in reply, 'Just as, before the birth of the founder of the school to which you belong, the theory it holds was not yet apparent as a sound theory, although it was really in existence, so likewise the opposite theory to that which you now propound may be already existent, though not yet apparent to us, so that we ought not as yet to yield assent to this theory which at the moment seems to be valid." (Sextus Empiricus 1990/second century, 26 or Book I, section XIII).

This argument must be recognized for what it is: it waves a white flag of surrender. We may suspend judgment on any dogma in the future. Therefore, we must suspend judgment on it now. How, then, do we learn anything philosophical at all? We do not! But we say we can because to deny it is to be refuted. The Pyrrhonian skeptics can keep up skeptical inquiry unceasingly because they will never be embarrassed by real knowledge.

VII

Richard Popkin (1960) remarked on the widespread influence of Sextus Empiricus' *Outlines of Pyrrhonism* after its Latin translation and publication in 1562. Academic skepticism had already influenced some theologians, like Desiderius Erasmus, a friend and critic of Martin Luther. That controversy raised the question of the criterion of knowledge and the rule of faith. The Pyrrhonian form of skepticism is much better suited to a traditional religious interpretation. It lends itself to a state of fideism or the doctrine that since we know nothing about ethics, we can rely on our orthodox faith to guide our lives. Custom, after all, guides a Pyrrhonian in practice.

But the custom in Europe was to follow the Catholic Church in Rome. We could argue against Protestant claims, which were not customary. They are, therefore, not sophisticated enough for the refined taste of the Pyrrhonian philosopher. Because of its use in religious debate in the Renaissance, Pyrrhonian skepticism became very widely known during and after the Reformation. By the beginning of the seventeenth century, it was no longer an obscure doctrine that concerned only a few isolated scholars of the Greek language. It had been translated and became a central issue in late Renaissance thought.

Francis Bacon was an Anglican. The theology of the Church of England in the late sixteenth century differed from the Protestant and the Catholic. The Church of England did not endorse the Protestant opinion that all knowledge sprang from the bible, *sola scriptura*. Nor did it support the Roman Catholic philosophers who conceded the futility of philosophy to the skeptic and asked us to rely on the sacred tradition of their Church for adjudicating our knowledge. The Church of England adopted some Catholic practices, and others were Protestant. But its philosophy was entirely beholden to neither. That ambivalent philosophy profoundly influenced Francis Bacon's *elenchus* and, indirectly, the shape of modern science.

Notes

1. Richard Bett (2000) makes a very good case that the "Pyrrhonian skepticism" of Sextus Empricus must be different from the kind of skepticism espoused by Pyrrho of Elis, if the latter was even a skeptic at all. It is also distinct from the skepticism propounded by Aenesidemus when he left the Academy, beven ased on the spotty record of his opinions. This would suggest that Sextus Empriricus was more original as a skeptic than often allowed.
2. For more information about the Hellenistic period of philosophy, the reader may consult Long and Sedley (1987).
3. Michael Frede (1987) brought to our attention the importance of the ancient study of medicine to philosophy and vice versa. Medicine was an early science to become autonomous, following Hippocrates, but like later sciences (physics, chemistry and biology for instance) philosophy always remains central to scientific thought, and vice versa.
4. See Owsei Temkin's Introduction to the *Gynaecology* (Soranus 1956).
5. Soranus (1956) seem to have a well rounded approach to medicine in his gyaecology, obstetrics and pediatrics. He was also quite knowledgeable about anatomy and physiology, though the Methodists have been criticisized for their neglect of the subjects. Perhaps the Methodists did not neglect them, but it is to be noted that their methods or their tests could be performed without taking account of anatomy or physiology when evaluating their results.

References

Bernard, Claude. (1864) 1957. *An Introduction to the Study of Experimental Medicine*. Tr. H.C. Greene. New York: Dover.

Bett, Richard. 2000. *Pyrrho, his Antecedents, and his Legacy*. Oxford: Oxford University Press.

Frede, Michael. 1987. *Essays in Ancient Philosophy*. Minneapolis: University of Minnesota Press.

Galen. 1985/2nd c. *On Sects for Beginners*. Tr. Michael Frede. In *Three Treatises On the Nature of Science*, ed. Richard Walzer, Michael Frede. Indianapolis: Hackett.

Long, A.A., and D.N. Sedley. 1987. *The Hellenistic Philosophers, Vol. I: Translations of the Texts*. Cambridge: Cambridge University Press.

Popkin, Richard. 1960. *The History of Scepticism from Erasmus to Descartes*. New York, N.Y.: Humanities Press.

Sextus Empiricus. 1990/2nd c. *Outlines of Pyrrhonism*. Tr. R.G. Bury. Amherst, NY: Prometheus Press.

Sextus Empiricus. 1995/2nd c. CE .*The Skeptic Way. Sextus Empiricus's Outlines of Pyrrhonism*. Tr. Benson Mates. New York, N.Y.: Oxford University Press.

Sextus Empiricus. 2000b/2nd c. CE. *Against the Ethicists*. Tr. Richard Bett. Oxford: Clarendon Press.

Soranus. 1956/2nd c. CE. *Gynaecology*. Tr. Owsei Temkin. Baltimore, MD. Johns Hopkins University Press.

5

Freedom by Confinement

I

Richard Popkin (1960) reviewed the years of the European Renaissance when ancient skeptical arguments became current again. They were deployed in a partisan manner in a central quarrel of the Reformation. The dispute concerned the "rule of faith," or the proper standard for determining our choice of religious affiliation. Faith and skepticism vied for attention when religious disputants conflicted over claims to know.

Augustine of Hippo was the most influential of all Christian philosophers. He was, like Cicero, a rhetorician. But the rhetorician Cicero had a great fondness for skepticism, perhaps to undermine his opponents less familiar with the use of critical arguments. He could use it in a law court and political discourse. Whether he was a skeptic beneath his rhetorical stance is not certain. Cicero's *Academica*, however, fully admitted that skeptics could not claim to know how to live.

A skeptic's *apraxia* appears undeniable. Of course, he could say that some opinions are more probable than others. But if he could not know the best way to live, he could not know that one way was *probably* best. Such a response, unfortunately, is not enough. Knowledge is dubitable in

J. Hattiangadi, *Francis Bacon's Skeptical Recipes for New Knowledge*,
https://doi.org/10.1007/978-3-031-52585-8_5

either case. An *elenchus* could refute but not affirm principles. In *Contra Academicos*, Augustine of Hippo ceased upon this weakness of Cicero's skepticism to invent a new form of religious opinion—a religion based on *faith*.

Augustine argued that since no skeptic could claim to know the truth about how to live, we must rely on blind faith to find our way in life. Faith is an opinion to which we feel loyal. It is ours, and we belong to it. Faith is so firmly held to be true that it does not allow for successful skeptical scrutiny. Religious faith differs from a philosophical dogma that may be revised upon discovering errors. Faith admits to no mistake. Once created, unshakeable faith became useful in a far-flung Roman empire that needed a simple claim to rally around.

Roman Catholicism and Orthodox Christianity adopted it, though they differed in the details of their faith. Some differences were there from the start between East and West. though a formal schism came several centuries after Rome became Christian. Christianity, however, became the faith around which subjects of the empire flocked, whether in Latin Rome or Greek Byzantium. The Nicene Creed, which Augustine, Bishop of Hippo, endorsed as faith, could no longer be questioned on skeptical grounds, as Arians had done. Nor could the doctrine of grace be questioned, as he outlined it, which was adopted in the Roman Latin or Western Christian tradition. The Eastern tradition was organized differently. In the Church, it was agreed on all sides to be orthodox, though they disagreed on what orthodoxy consists.

Only after Thomas Aquinas had updated Church theology was the overwhelming influence on Church doctrine of the philosophy of Augustine of Hippo moderated. Human rationality was allowed to play a larger part in determining knowledge, always provided that it was compatible with the Gospel of Christianity. The Western Church began to regard the relation of faith to reason more favorably. They were thought to work together in interpreting the bible. The Roman Catholic Church continued to reserve the authority to interpret the bible. Orthodoxy remained. Collectively, the bishops decided on questions of biblical interpretation, balancing faith and reason—based on their faith. Outside the episcopate, i.e., beyond the collected bishops, others could also employ reason to inquire about things—provided that they conformed to official orthodoxy. On key questions roiling the Church, the Pope would announce formal decisions from time to time.

From the time of its founding by Peter the apostle, the Church had helped interpret the bible to the flock. The scope of reason beyond theological orthodoxy was always a question of controversy. With the Reformation, there was a return to some of Augustine's ideas. Reformers accepted the bible itself as the sole criterion of truth about faith. Martin Luther refused to allow that the sacred tradition, or the Church, is required to determine any interpretation of the word of God.

According to all Christians, knowledge gained from the bible was relevant to faith, but Martin Luther claimed it is all we need. The bible would determine its own meaning. The principle was "*sola scriptura*." He objected to using Thomas Aquinas' or any other philosophy to interpret Biblical passages. The reformers argued that Thomist thought is merely another human opinion and should not be made binding on Christians everywhere. We have the divine bible, and it is enough.

Skeptical arguments came naturally into the mix in the sixteenth century. The ancient skeptics had raised the question of the criterion as a rule of knowledge in a general way. The Church could respond to the Reformers that we need a criterion to determine which interpretation of the bible is correct when there is a disagreement among serious theologians. Such an argument challenges *sola scriptura*. Protestant reliance on biblical knowledge can be challenged as founded on human opinion—even if the bible is divine. The entire array of skeptical arguments against a criterion of knowledge can be deployed against the Reformed rule of faith. Desiderius Erasmus was the first to challenge the view that anyone can know what scripture says if they merely read it, even with the purest of hearts.

The Church claimed to be the divinely authorized institution, embodying the sacred tradition of the episcopate, and it would appear to be just what we need to show us the way. It was, after all, set up by the Apostles, and they were close to Jesus Christ, and blessed in person. Therefore, The episcopate was the proper tradition to interpret scripture when most needed. They were using arguments from the skeptic's arsenal to undermine reformers during the counter-Reformation (Popkin 1960).

Skeptical arguments are corrosive to all that touch them, however. Whenever counter-reformers claim that reformers lack an adequate criterion of knowledge because a criterion of *sola scriptura* is circular, it can rebound on them. To be able to defend their own faith, Roman Catholics found it necessary to defend the interpretation of the bible on the basis of the sacred tradition. There is a difficulty for them too. We need only ask whether what the sacred tradition tells us about itself is as reliable as the bible. If we cannot interpret the word of God because there are many possible interpretations, how can we interpret the word of mere humans who claim to be the authorized interpreters of the word of God? The skeptical difficulty of the criterion of knowledge reappears. There can be different opinions as to what tradition implies for scripture. The opinion within the so-called sacred tradition is only one among many.

The reformers offered the doctrine of *sola scriptura* because when Christians become true converts, they do not judge the truth of scripture themselves. At the moment of conversion, the person is possessed by the holy spirit, a true angel, and therefore acquires a kind of divine knowledge of divine scripture. What could be better? There remains a difficulty for that claim, too: they need to know first what possessed them before they can trust their enthusiasm.

The reliance of the sacred tradition on skepticism to defeat reform has its own problems. Is a skeptical Christian truly Christian? The argument is this: A Roman Catholic who is also a skeptic, like Michel de Montaigne, can defend his faith as true because it is customary. Custom is all that remains once we recognize that there is no criterion for determining the truth of any religious knowledge. For if we accept a criterion on its own grounds then it is circular, but if we need another criterion for the purpose, we are led to infinite regress. So a skeptic falls back on custom or tradition, which happened to be the Roman Catholic or Latin tradition in Western Europe. But this creates its own difficulty.

Is a skeptical Christian truly Christian? Being a skeptic, he cannot assent to a doctrine like the depravity of Adam's progeny when expelled from Eden. Original sin may be professed, but is it believed? He cannot believe that God alone can give us our moral compass and that the scripture alone can save us from mortality and sin. Of course, if everyone around him accepts these doctrines, the skeptical Christian will be willing to live in conformity with the customary ethics. But he has no reason to believe any of it and every reason in the ample arsenal of Sextus Emiricus' Pyrrhonism to refuse to assent. If he does not assent to God's commandments, then he can live the depraved life of a "libertin érudit" (supposed to be depraved, anyway) while superficially conforming to outward Christian custom.

The skeptical double-edged sword led the "Pyrrhonian Crisis of Faith." It formed a backdrop to the development of modern philosophy in Europe. Academic skepticism and forms of fideism nourished by it were already known in the period between 1520 and 1560. Desiderius Erasmus, Thomas More, Cornelius Agrippa von Nettesheim, and others wrote about the futility of using reason in such matters. Michel de Montaigne's *Essays*, especially the long 12th chapter, "Apology of Raymond Sebond," made the fideist case more widely known. Although earlier authors may have relied more on Cicero and Diogenes Laertius to understand skeptical arguments, it seems that Michel de Montaigne relied more on the text of Sextus Empiricus. At any rate, the skeptical crisis of the late sixteenth century may justly be called the Pyrrhonian Crisis, as Richard Popkin notes, and it was widely discussed as such in the years following 1562 when the Latin translation of the *Outlines* was published (Sextus 1990/2ndc.) (Popkin 1960).

Natural Philosophy was not spared in the theological debate. Natural science, as we now call it, had to keep clear of the unorthodox in its claims. Science was subject to theological review. The method of subjugation differed. In the Roman Catholic tradition, it was subject to review by the Censors of the Inquisition, the original thought police. In the Reformed Churches, it was held that all knowledge flows from scripture. Christian science sets the limit of legitimate science. The Elders of the Church know, as do all those of good conscience able to read the bible—and all others who disagree are unfortunate and depraved. Orthodoxy had, in any case, strangulated free thought.

Francis Bacon did not accept that there is a clash between reason and faith. He could rescue science or philosophical knowledge because he did not countenance such a clash. If he was troubled, he wrote nothing about it. But his writing does show a way of avoiding it. He may have reflected on it early, perhaps as early as 1685 (see Matthews 2008). He found a reconciliation of faith and secular reason in Anglican theology by finding a legitimate use for skeptical reason within the bounds of faith.

Francis Bacon achieved the reconciliation by refusing to let reason adjudicate questions of faith. Belief is more important than knowledge in any faith-based account of knowledge. Skeptical reason is then confined to studying knowledge within strict bounds determined by such a faith. Such was the state of philosophy, whether in Catholic or Protestant jurisdictions. Francis Bacon altered this relation. The bounds of reason were defined within his faith and by his faith. But once the boundary is set between secular study and faith, skeptical doubt can roam freely within secular studies, such as natural philosophy. Faith will reciprocally not adjudicate on matters of reason.[1]

Francis Bacon's philosophy went beyond the warring factions of the Reformation to give the philosopher a new freedom to think. His confinement of reason and faith was different. He confined religion within its bounds. It remains unchallenged within those confines. But beyond that confinement, knowledge is free and can engage in skeptical questioning with abandon—so long as the questions do not infringe on faith.

The sharp separation he contemplated between the two activities of the understanding was there to protect faith from reason. At the same time, it allowed reason, skepticism, and modern secular science to be free from religious scrutiny. Good fences make good neighbors. Boundaries may be used to protect both sides if disputes arise. Of course, Francis Bacon's idea of a skeptical method owes much to Anglicanism, which allowed for such boundaries. Modern secular science owes its freedom from religion to the space it was given within the Church of England. Such is the paradoxical origin of secular modern thought in Europe.

Francis Bacon's basic thought is expressed succinctly in the opening section of his early and unfinished book, *Valerius Terminus of the Interpretation of Nature with Annotations of Hermes Stella*. After abandoning this book, he published his *Proficiency and Advancement of Learning*, which carries the same message in greater detail (Bacon 2000/1605). His story is biblical in origin. God made the world as he did and gave his human creatures, dominion over it. With the Fall, described in the bible, humans lost their dominion on earth. Driven out of paradise for the sin of hubris, they became depraved. They needed divine assistance to live a good life and avoid evil. Human beings were floundering, not knowing how to live. God took pity on humans and sent messengers to tell us his will on how we must live. Scripture, consequently, is a source of direct knowledge of human ethics, revealing God's will.

How God uses his power in the natural world is different. Nature remains obscure. It is not part of ethics, nor spelled out anywhere in the bible. Faith cannot yield knowledge of the natural world. However, reason, or *logos*, can help us there, which God has also provided us. Baconian and Thomist opinions coincide in this respect. The bible does not give us the key to his creation, only to his will as to how we should act. If we want to regain our dominion over nature, we must use our faculty of reason to fathom the world's secular secrets. He has given us *logos*, and *dialectic*, but we have to learn how to use them.

Francis Bacon looked for the right biblical way to reason in natural philosophy. That is how secular science has a rebirth in Europe. "Moses… is said to have been seen in all the Egyptian learning, which nation was early and leading in matters of knowledge. And Salomon the King, as out of a branch of his wisdom extraordinarily petitioned and granted from God, is said to have written a natural history of all that is green from the cedar to the moss… and also of all that liveth and moveth. And if the book of Job be turned over, it will be found to have much aspersion of natural philosophy. Nay, the same Salomon the King affirmeth directly that the Glory of God IS TO CONCEAL A THING, BUT THE GLORY OF THE KING IS TO FIND IT OUT…." (Bacon 1857 v, 3). His use of capital letters perhaps signify the importance of the ideas he is drawing to our attention. Whereas divine faith is transparent, natural knowledge is opaque. Unlike God's law, nature remains intriguing, and he sets out to let us see through its veil (Klein and Giglioni 2020).

II

Limiting the Church arose as a political necessity in Elizabethan times. Anglicans repudiated the Protestant doctrine that the bible interprets itself, a form of the principle of *sola scriptura*. But they also abandoned the full Catholic tradition as practiced in Rome, asserting it is enough to believe, *sola fide*. The Church of England asserted the legitimacy of its version of the sacred tradition. Anglican bishops traced their beginning to Peter, the Apostle, and were therefore also blessed, like those in Rome. The Church of England accommodated both some Catholic sacraments and some Puritan reforms within a broad conception of the Church. There were some issues over which there was a bitter fight between factions of the Church, such as the book of common prayer, the vestments, and the status of the Apocrypha. But it kept a balance between the two warring sides. It was Protestant but not Puritan. Francis Bacon worshipped in this finely balanced Church—a Church hesitant to assert controversial theological doctrines. To keep the peace, its theological claims were kept to a minimum.

The reconciliation of faith and skeptical reason by sequestrating faith is an interesting way to approach the study of natural philosophy. Francis Bacon did not agree with the Protestants that the scripture is a self-authenticating document because he must allow, as an Anglican, that the Bishops in England are needed for its proper reading. But he could not agree with a blanket approval of tradition, based on Pyrrhonian doubt, because the customary Roman tradition had been repudiated in England. England had a bespoke English faith. Francis Bacon, therefore, relied on a new reading of the Fall.

His main innovation was how he related the study of nature to the need for *philanthropy* or *charity*. Truth is not to be pursued for its own sake in natural philosophy but only to discover valuable experimental techniques when they are of benefit to humans. Skeptical science is to be pursued under God's injunction to be charitable—to extend Christian love to all unfortunate humans. This love depends upon humans regaining their lost dominion over nature, which is their right. They had lost it from too much hubris. Now charitable love will bring them back their power over nature. Showing how skepticism can lead to charity was the task of the Baconian method.

Francis Bacon's reconciliation did not depend upon any particular rule of faith in interpreting the scripture, whether Puritan or Catholic (Matthews 2008). Whether scripture and Church ritual are interpreted by tradition or by conscience is undecided in his writing. He left it to the Bishops to decide such issues. Francis Bacon asks philosophers to avoid such questions. Faith is shielded from skeptical scrutiny in his account by forbidding the search for knowledge in the circle drawn around faith. We cannot investigate God's will, as Adam tried in his moment of hubris. By restraining our philosophy, we liberate natural philosophy from religion. We usher in modern secular science. He rediscovered the freedom of Plato's pagan philosophizing entirely within the Anglican faith. That was how he differed from the other reconciliations of reason and faith in his time. However, once he had released the pagan genie from the bottle, there was no putting it back.

Adam, by his Fall, had lost dominion over nature. He had it once, and Francis Bacon decided he knew how to restore that dominion. *The Great Instauration* (or the Great Renewal) was his name for a future six-part plan to restore it. The first part was to review all received knowledge of the sciences and the arts. He attempts this in his 1605 book on *The Proficience and Advancement of Learning, Divine and Human*. It is a curious work, giving us a brief glimpse of a future scientific edifice. He explains that this edifice must be built on experiments. However, the results of experiments are not predictable. What he provides in that book is, therefore, provisional at best. He sketches an overall plan, withholding key details of how it will achieve its task.

It reads like a grant application for a research project. Applying for support may have been part of its intent. He wanted royal financial support for the equivalent of a government department to pursue modern science. But the book is too long for that effort alone when you compare it with the two-page letter of supplication of 1592 to Lord Burghley. That letter was a grant application. There is no reply on record from his influential uncle. It was probably quietly but firmly rejected.

Francis Bacon's contemporaries may have felt a need to defend learning from the Pyrrhonian attack. He did not. The first part of his grand six-part plan for philosophy was to review and reclassify the stock of all knowledge we possessed before his time. *The Proficience and Advancement and of Learning, Divine and Human* began that review process. At the very least, the *Advancement* affirms our need for older science. Contrary to Pyrrhonism, we should not live merely by appearances alone. We must seek reality and affirm it when we have it because new knowledge gives us new power over nature. New powers bring added benefit to humans. The divine command to be charitable has no prescribed limit. Our duty of charity requires us to pursue new knowledge to alleviate suffering. *Ataraxia* is not enough.

Francis Bacon wrote *Advancement* to convince everyone that faith need not fear the Great Instauration. Its skepticism was benign, he averred. The danger that skepticism posed to faith was on many minds then. It is understandable that Francis Bacon tackles its containment first and tells us the details of the new skeptical method some fifteen years later. Nevertheless, Francis Bacon does not say as much about skepticism as he could have said to clarify his message. His reticence is not surprising: he must not alarm the faithful. Orthodox Christian thought since Augustine of Hippo required the faithful to abjure skepticism. A full-fledged statement of the use of skepticial methods would have frightened his audience—even if he promised that faith would be duly caged and thus protected from its challenge. The Pyrrhonian Crisis was raging all around him. His reticence may have been wise at the time. But it almost succeeded in hiding his meaning and his method altogether—not just from all but a few of his contemporaries, but also from posterity.

III

In the writings of Plato and Aristotle, several explanations were found to explain why any event occurs or why anything is just what it is. "Becauses," or explanatory causes, were well-rehearsed in scholastic teaching. There are, first of all, the formal and the material causes. Suppose we want to understand why a sculpted figure is what it is. The matter, e.g., clay or stone, which is molded or chiseled into shape, illustrates the cause called "material." Its eventual stable shape presents the formal cause of the figure. Every figure will have these two causes: its form and matter. Let us focus on how the sculptor molds or chisels the figure. We can discern two more becauses, one being the mechanical cause, or the action of the chisel on stone, or the hand on the wet clay. The other is a final cause, illustrated by the sculptor's end or purpose of producing the sculpture. This fourfold classification of ancient origin was clearly a well-thought-out scheme that had stood the test of time—for over a millennium and a half. Most of his scholastic contemporaries thought we need to understand all four causes to understand the nature of anything.

In the *Timaeus*, however, Plato had urged us to consider a divine creator of the cosmos, a demiurge, to understand natural causes. Aristotle followed Plato part way in this line of thinking. There are subtle differences, but the differences bear on matters of faith. Francis Bacon could not allow differences between natural philosophers to influence matters of faith. Philosophers must not dabble in the purpose of creation. That is where faith rules.

Francis Bacon objects to Plato's meddling in matters of divine purpose in pursuing natural philosophy. We must keep the skeptical impulse out of theology to protect the faith. For this reason, he finds that the pre-Socratic Philosophers were better as natural philosophers than either Plato or Aristotle or the schools of thought they inspired (neo-Platonism, and scholasticism, respectively.) Alchemists who combined neo-Platonism, astrology, number mysticism, and celestial magic in their philosophy were abhorrent to Francis Bacon because they, too, crossed the line. Paracelsus is dangerous because he brought theology into his alchemical and medical thought. He even tried to give his bewildered reader a scriptural basis for alchemical opinion. All these attempts violate the separation of skeptical science from religious faith. They filled Francis Bacon with horror.

Francis Bacon went further than disallowing scriptural arguments in natural philosophy. He also disallowed the study of final causes (goals or ends) within the study of physics. Material and mechanical causes seem to him to be sufficient to understand the natural world at the basic level. In isolation, such abstention looks very modern. His reasons were not secular, but religious. The effect, however, was a new secular science of nature.

He speaks of Form even in the physical realm, but his conception of Form is not the Idea in Plato, but modified. It is divorced from its theological use in ancient thought.[2] He says we should emulate "the school of Democritus, which went further than the rest. Matter rather than forms should be the object of our attention, its configuration and configuration changes, simple action, and the law of action or motion; forms are figments of the human mind, unless you will call those laws of action forms." His thoroughly modern understanding of the world in terms of matter and mechanical at the lowest level of knowledge is derived not from an uncanny look into the future of science but from his attempt to integrate faith with skeptical reason without endangering the faith.

Francis Bacon describes his new science as structured like a pyramid, with "physic" at the lower and "metaphysic" at the higher end. When we reach the topmost regions, we get to formal and final causes properly understood (Bacon I.17–19 and 104). When we are near the apex of the edifice we are dangerously close to theology. The apex therefore belongs to the Church. As the Church's supreme governor, the King must take over its control. Only the Church of England has jurisdiction on that little pyramid crowning the rest of the pyramid of knowledge beneath.

Francis Bacon valued any experimental knowledge gained by trade practitioners (including alchemists.) He directed his thoughts of the new science to improve on their "experiments." Alchemists were great experimenters. But the results of their experiments were trade secrets. They built their trade by exploiting their experimental proficiency. Unlike the trades, the new Baconian science would assume the form of *public* knowledge. Unlike alchemical knowledge, it is emphatically not a secretly coded publication. It should not be written to be understood only by the initiates within a guild or a trade. Charity and not profit must motivate science. Charity was to remain a defining mark of pure science in the centuries to follow: fundamental research remains a public enterprise.

Francis Bacon proposed a public science that was experimental. Experiments would somehow yield us new power over nature, with the promise of practical benefit to those in need. We can pursue natural knowledge as part of the Christian duty of charity. The publication of scientific results was for Francis Bacon, and later for Robert Boyle, a Christian duty and not an act of self-promotion.

Following Francis Bacon's conception of a charitable science, a public science, Robert Boyle later required that all experimental results of the Royal Society be published. In recent years the general public character of science has often been misunderstood. It has been confused with whether its experiments are conducted in a publicly funded or privately owned space. We imagine that research conducted in a private space would be biased or contaminated by the owner's financial, or political interests. Baconian science was not conceived to be free from commercial interests. Francis Bacon lived in a partly feudal age and seemed not to have noticed the growing mercantilism around him. The profit motive was not an overwhelming issue that arose then, though it may today. The purpose of publication, i.e., to make any knowledge public, is to facilitate the dissemination of ideas. Free dissemination would speed up the new practical benefits to the poor and needy. The duty of charity required it.

Private developments of applications of public science are now common. They have increased in number over the years. Applications of the Baconian method have thus been used for profit. A dramatic growth in patent letters since the seventeenth century is a measure of the usefulness of modern science. New knowledge can provide immense profits now. Private profits may be so robust that they may even sometimes fund research in fundamental (usually public) science. The profit motive of those supporting research may well direct the research they support towards more private profit. Such a distortion of public research is a matter of genuine concern today. But we cannot understand Baconian science by applying our experience from this later economic development.

IV

In Book II of the *Novum Organum*, Francis Bacon announces his task thus: "On a given body, to generate and superimpose a new nature or new natures is the task and aim of human power. Of a given nature, to discover the form, or true specific difference, or nature engendering nature, or source of emanation (for these are terms that come nearest the description of the thing), is the work and aim of human knowledge." (Bacon II.1)

There is a striking juxtaposition of works and knowledge. He puts the task of producing works first. Works come before human wisdom. The applicability of wisdom is its purpose.

While still a young Cambridge University student, Francis Bacon studied scholastic reformulations of Aristotle's works. He developed an aversion to scholasticism, if only because the scholastics tried to interpret scripture according to philosophical reason—as shown, for instance, in the work of Thomas Aquinas. But Francis Bacon rejected all of this, as had the Calvinists and Lutherans, though he had different reasons. He refused to endorse the thought of Paracelsus von Hohenheim, who tried to base his oracular ("auricular") sayings in an interpretation of scripture.

He calls the scholastics and the alchemists the two rogues ("rovers") of philosophy. He owes much to these two "rovers," as he calls them, for his ideas on method, perhaps because he rejected each.

The rejection of ancient and contemporary schools of thought came easily to him. Francis Bacon was, after all, a skeptic. But he was undoubtedly also fashioning his writing on the skeptical rejection of schools of thought, for instance, in Michel de Montaigne's writings, who was also a skeptic.[3] No school of thought could obtain their full assent. Francis Bacon's writing is full of rejections of older and contemporary philosophies, even though he thought we could learn from them when properly reformulated. He thought we should reserve our assent so that the eventual outcome of our search for reality will give us the truth, profitable inventions and new discoveries, all at once.

Apart from the aim of discovery, which is charity, Francis Bacon's thoughts are a bit like Plato's, who regarded the search for true answers as superior to persuasive ones. Appropriately, Samuel Taylor Coleridge called him the British Plato (Coleridge 1859/1818). Other commentators have often understood him as a radical who wanted to begin philosophy anew out of dissatisfaction with what went before. However, he was not that radical. Francis Bacon was a great admirer of Graeco-Roman life. He thought that there was a great deal to learn from them. In his book, *The Wisdom of the Ancients (Sapienta Veterum)* he sets out to salvage some of the old wisdom that he thinks we should preserve and build upon. In it, he recounts the wisdom of the past as parables about ancient myths. He was a skeptic who valued existing knowledge claims. There is no paradox here. He did not recommend that his new science was to be carried out by those who lacked learning. He hoped they were not gullible enough to assent to earlier schools of thought. Learned ignorance, to use a phrase from Nicholas of Cusa, is what Francis Bacon sought. To aid them in their pursuit of the new philosophy, he warns them that the mind has certain phantoms that distort our ability to learn, called "idols of the mind." (Bacon I.38 and 115) Idols may distort ancient learning, but the learning remains useful, all the same, he thought.

The word "idols" is more commonly used as a physical representation of divinities, as in temples. Such is not its meaning here. Francis Bacon's use of the word is more in line with the word "eidola" used in Plato's myth of the cave, in his *Republic*, and in Plato's *Sophist* (Zagorin 1998). *Eidola* are described there as phantoms of the mind, as distortions of reality. Bacon's Idols of the Mind are also false images that distort reality, far removed from the true Ideas or the Forms of things in the mind of God.

He classified Eidola under four headings. There are the Idols of the Tribe, which are those false ideas that our mind has by virtue of our being of the tribe of humans. These are distortions in our understanding by our nature, which we cannot escape.

There are also the idols of the cave, so-called in reference explicitly to Plato's allegory of the cave. These idols are individual shortcomings that may interfere with our ability to see things as they are.

The third division is called in different texts Idols of the Market Place, of the Forum, or of the Palace. If he meant the same thing by them all, then in this category of idols are those that arise from what we call society. Language is most important. He included in it the common words we use to describe things, which cloud our judgment when we try to discern what and how things indeed are. He thought that our terms do not cut reality into kinds in the same way as reality is divided.

Only these three kinds are listed in an early and unpublished work, but in most of his writing, a fourth is also listed, namely, "Idols of the Theatre." He is thinking here of the opposing schools of philosophers who debate with one another on the intellectual stage. We are inclined to assent to one or another opinion as true. To do so, however, is to fall prey to this distortion of the understanding that interferes with our ability to search for the true nature of anything. It is also a pernicious Idol.

In his early and unfinished piece, "The Masculine Birth of Time," he also mentions two other Idols that he contrasts with one another: the "Idols of the Inn" and "the Idols of the Road." The first are the idols that you face when your mind is at rest and beset by the usual Idols, but if you take the existence of these Idols to heart and try to run away from them, he thinks you only incur new Idols, namely the Idols of the Road. There is nowhere to turn from the Idols. One is never free of Idols. His point of view remains skeptical, allowing no means to learn "kataleptically." Grasping the true and full reality of things themselves is not within our reach—not that is, without learning his new method.

One startling omission is this: there are no Idols of faith. If he was an Anglican, presumably, there are Roman Catholics, Calvinists, Lutherans, not to mention Muslims, Jews, Hindus, Aztecs, Confucians, and Zoroastrians… who would think of the world differently. Presumably, they cannot all be right. All but one fail, and perhaps not even one is true. Faith can also distort one's opinion. But faith is missing in each of Francis Bacon's many lists of Idols.

V

The first summary of his new method on record is in his 1592 letter to his uncle, William Cecil Burghley. Francis Bacon summarizes the two philosophies he hopes to improve upon (Spedding 1861–74). Their demise is the basis upon which his new method will be built. In all later writings, he takes pains to reject every school of thought that occurs to him, whether ancient or contemporary. His negative remarks about the schools make it confusing to the reader. Is he a radical, or is he not? But at this early stage, he lists two for particular censure: scholasticism and alchemy. They are rejected but retain a specific relevance to his new method.

He objects to the scholastic method, which was derived mainly from Aristotle. Scholasticism depended on observing things under normal conditions to discover their true nature. The nature of an item is stated as a definition derived from everyday observations of it. We must abstract from our perceptions of it to get its essence. Once we get the definition of its essence, syllogisms and logic are used to derive conclusions from the definition. The definition is used as one of the premises for an understanding of natural phenomena. He found this method too limiting.

Francis Bacon was interested in "works." He lists three great inventions of his time as impacting human well-being: the discovery of magnets, the discovery of gunpowder, and the printing press. The magnetic needle had transformed navigation for the better, gunpowder had transformed the art of warfare, and the printing press had transformed the dissemination of thought. None of these inventions could have been derived from or inspired by scholastic principles. Which nature would we study to make such inventions?

Take gunpowder, for instance. When charcoal, saltpeter, and sulfur are mixed in the right proportion, originally as 10, 3, and 1 units, respectively, the black powder forms a highly explosive mixture. It can propel a projectile at immense speed over a great distance and thus becomes useful as a new weapon of warfare. But if we study the nature of these three earths from an Aristotelian understanding of first principles, we would not find the formula for gunpowder. The nature of gunpowder cannot be derived logically from the scholastic study of the nature of charcoal, saltpeter, and sulfur. What we need to do is to experiment. By experiment alone, alchemical artisans had discovered that the mixture of these three earths is explosive.

Let us turn to alchemy. It had a venerable ancestry going back to the East, as far as Egypt, India, and China. Gunpowder was formulated first in China, though its use in warfare was nurtured in Europe and West Asia. We find it practiced by people seeking practical knowledge. The transmutation of base metals into gold and the search for an elixir of life are practical ends rather than contemplative of nature. As practiced by glassmakers in Egypt, for instance, alchemy was a subject steeped in secrecy. Knowledge was kept within the guild to protect trade secrets. Alchemists sought to extract income from their exclusive experimental understanding. Their practical knowledge was also valuable for metalworking, mining, inks and dyes, producing liquors and spirits, etc. Apart from their efforts to keep their techniques secret while writing instructions for their use in their trade, alchemical texts are hard to decipher for other reasons. The alchemical theory was also tied to astrology, numerology, gnosticism, cabala, and hermeticism. Its Hermetic imagery has not yet been fully deciphered. The alchemists' practical techniques were hidden in obscure astrological formulae that prevented easy reading. Such was not the charitable, public science that Francis Bacon wanted.

Alchemy gave us isolated experiments, useful ones, no doubt. But, they failed to add up to a system of knowledge in which one can understand why the experiments work. Whatever theory we are given serves only to hide and obfuscate what we know. The effort to explain and make it clear is missing. Profit was all. Therefore, Francis Bacon made a fundamental distinction between two varieties of experiments: fructiferous and luciferous. Fruitful instances are many, but they are isolated because we cannot discover more experiments by learning from the fruits of one. Luciferous experiments, however, shed light on the nature of things.[4] They give us knowledge. They will provide insight into true natures and, thus new and useful experiments that are hitherto unknown. While fructiferous experiments that are hit upon by accident are valuable for what they yield, better value lies in a method of science that can generate new luciferous experiments (Bacon I.99). Alchemists, seeking profit, were only interested in fruitful experiments and did not know how to seek enlightenment.

Skeptically, he sets each school of thought against the other. Of scholastics, he says that though they explain everyday reality, they fail to give us valuable experiments. Of the alchemists, he says that though they give us some beneficial new experiments, we do not grasp the natures of things in alchemy well enough to understand their everyday reality. Lacking an understanding of the nature of things, we cannot utilize our practical alchemical knowledge to generate new and useful experiments. Philosophy is barren of works, while alchemy is barren of understanding. Hence, describing his new method as fashioning "experimental and natural histories" was to combine these two virtues.

Using the need to produce works as a criterion of adequacy for any future science, Francis Bacon found both the scholastic theory of matter and alchemy wanting but useful. He showed how to use the proficiency of existing learning to lead to its advancement. He asks us to reject them but is fashioning a new method based on their parts. One part is the public and everyday knowledge of the world, as found in scholastic observations and natural histories. Combined with that part is the dark and mysterious art of experiment, practiced by alchemists. Alchemy challenged the observed nature of things. His combination gives us his new method that mixes observations and experiments to investigate nature through cross-examination.[5] His unique method is a thoroughly skeptical examination. We use experiments to question the results of observations and vice versa—a cross-examination of nature, much like a Socratic dialogue about what is known.

It is difficult to say which of these came first: his clever idea of a skeptical examination of nature by observation and experiment or his technique of insulating his faith from skepticism. We can make a plausible account for each case. If he had discovered his clever technique of interrogating nature first, he would have been obliged to consider how skepticism affects his faith, given the Pyrrhonian Crisis that was roiling contemporaries. On the other hand, he may have discovered his compatibilist solution to the Pyrrhonian Crisis from the prevailing Anglican ambivalence about Church dogmas. And he may then have developed his clever idea of cross-examining the evidence to restore our dominion over nature.

Notes

1. Peter Harrison has made excellent arguments from time to time about the theological background of modern science. For instance, his (2023) is a general summary of his opinions on this subject. While there may be differences in matters of detail in this chapter, particularly concerning the interpretation of Anglican thought in Francis Bacon's time, there remains a major difference of reading of modern science. In this chapter, modern science is interpreted as fully secular, and as pagan as Plato's philosophy. Theology is no doubt still important to study to understand the emancipation of modern science. We need to study the theological background of modern science to discern how it escaped being submissive to theological orthodoxy, as it has.
2. When or Samuel Taylor Coleridge (1818) calls Francis Bacon "the British Plato" or Harvey Wheeler (2001) notes Plalto's influence, they are right to see the connection. But it is also important to note that while Plato's skepticism is adopted fully, his doctrines are accepted only in a very modified form.
3. Dana Jalobeanu (2019) has put together many of Francis Bacon's scattered remarks on the ancient skeptics and shows that his comments reflect the influence of Michel de Montaigne.
4. Francis Bacon has been assigned by historians to a mysterious and well-hidden "maker's knowledge tradition" (Pérez-Ramos 1988; Henry 2002), in which a maker is thought to have complete knowledge of the nature of the product made. If there is such a tradition, it would seem that he did not belong to it. Fructiferous experiments, which craftspeople can use make things, are not described as enlightening enough to substitute for the luciferous experiments needed in science.

5. Brian Baigrie asked in 1998 why boxes and bins of manuscripts in late seventeenth century science were marked "Observations and Experiments" and never just one or the other. It is in Francis Bacon's method that the answer is to be found. His new method is a skeptical combination of the two methods of ancient origin.

References

Bacon, Francis. (1603) 1857. Valerius Terminus of the Interpretation of Nature: With the Annotations of Hermes Stella. In Vol. 3 *of The Works of Francis Bacon*, ed. James Spedding, Robert Leslie Ellis, Douglas Denon Heath. London: Longman & Co.

———. (1605) 2000. The Advancement of Learning. In *The Advancement of Learning*, ed. Michael Kiernan. v. IV, *The Oxford Francis Bacon*. Oxford: Clarendon Press.

Coleridge, Samuel Taylor. 1818. General Introduction: Or Prlimary Treatise on Method. *Encylopaedia Metropolitana*. v.1 (1–43). ed. Edward Smedley, Hugh J. Rose, Henry J. Rose. London: Fellowes and Rivington.

Harrison, Peter. 2023. Normativity and the Critical Function of Genealogy: The Case of Modern Science. *Modern Theology* 39 (4): 682–707. https://doi.org/10.1111/moth.12857.

Henry, John. 2002. *Knowledge Is Power. Francis Bacon and the Method of Science*. Cambridge: Icon Books.

Jalobeanu, Dana. 2019. Francis Bacon on Sophists, Poets and Other Forms of Self-Deceit (Or, What Can an Experimental Philosopher Learn from a Theory Informed History of Philosophy?) *Experiment, Speculation and Religion in Early Modern Philosophy*, ed. A. Vanzo, and P.R. Anstey, New York, NY: Routledge.

Klein, Jürgen, Guido Giglioni. (2020). 'Francis Bacon', *The Stanford Encyclopedia of Philosophy* (Fall 2020 Edition), Edward N. Zalta (ed.), https://plato.stanford.edu/archives/fall2020/entries/francis-bacon/.

Matthews, Steven. 2008. *Theology and Science in the Thought of Francis Bacon*. London: Routledge.

Pérez-Ramos, Antonio. 1988. *Francis Bacon's Idea of Science and the Maker's Knowledge Tradition*. Oxford: Clarendon Press.

Popkin, Richard. 1960. *The History of Scepticism from Erasmus to Descartes*. New York, N.Y.: Humanities Press.

Sextus Empiricus. 1990/2nd c. *Outlines of Pyrrhonism*. Tr. R.G. Bury. Amherst, NY: Prometheus Press.

Spedding, James, ed. 1861–74. *The Letters and the Life of Francis Bacon*, 7 vols. London: Longman, Green, & Co.

Wheeler, Harvey. 2001. The Semiosis of Francis Bacon's Scientific Empiricism. *Semiotica* 201 (133): 45–67.

Zagorin, Perez. 1998. *Francis Bacon* Princeton: Princeton University Press.

6

The Sphinx

I

The following passage from Francis Bacon's *New Organon* has been quoted often. He speaks obliquely of science as neither a mere collection of facts, as in ancient empiricism, nor the method of hypothesis, as in ancient rationalism. Science, he says, does better than both. It is up to us to decipher what his third path does.

> Those who have handled sciences have been either men of experiment or men of dogmas. The men of experiment are like the ant, they only collect and use. The reasoners resemble spiders, who make cobwebs out of their own substance. But the bee takes a middle course: it gathers its material from the flowers of the garden and of the field, but transforms and digests it by a power of its own. Not unlike this is the true business of philosophy; for it neither relies solely on the powers of the mind, nor does it take matter which it gathers from natural history and mechanical experiments and lay it up in memory whole, as it finds it, but lays it up in the understanding altered and digested. Therefore from a closer and purer league between these two faculties, the experimental and the rational (such as never yet been made) much may be hoped. (Bacon, I.95)

J. Hattiangadi, *Francis Bacon's Skeptical Recipes for New Knowledge*,
https://doi.org/10.1007/978-3-031-52585-8_6

We know that his method is a ladder of ascent to form a pyramid-like structure for knowledge. At base, it begins with a natural history, collected using a Socratic cross-examination of nature. But where do we go from there? One of the most challenging things in understanding Francis Bacon's writing is what he calls the "helps" he provides (Bacon, II.19–51). When we examine what helps he offers us, it turns out to be one of two things: he provides us a method of using experiments to undermine observations and each other, and asks us to make tables out of these instances. Having illustrated these tables in the latter part of the *Novum Organum*, he lists topics he hoped to pursue but stops short there. Instead of new science, he embarks on summaries of other parts of his six-part work, of which the unfinished *Novum Organum* was only the second.

The titles he provides us and abandons are of little help to decipher what Francis Bacon had in mind as his method. How do we digest the information and convert it to something useful, as does the bee? He lists the following: "Prerogative Instances; Supports of Induction; Rectification of Induction; Varying the investigation according to the Nature of the Subject; Prerogative Natures with respect to the Investigation (i.e. the order of investigating natures); the Limits of Investigation; Application to Practice; Preparations for Investigations; and lastly, Ascending and Descending Scale of Axioms." (Bacon II.22). While we could interpret this list in terms of what else he says about such things, this list does not shed much light on the rest of his inductive method.

He may have found that what he had to say did not amount to much. Such a judgment may well have been why he abandoned more elaboration along these lines. Nevertheless, what this tells us is meager, and we can get it from his other writing. He tells us that induction is not unerring. It needs rectification and adjustment. Somehow, it leads eventually to a ladder or scale of axioms. The levels ascend to the highest level, which must be reconciled with lower-level discrepancies to make them work. These mutually reconciled axioms give us abstract knowledge and help us find practical applications when we descend using logic to the level of lower axioms. None of these seem to be the kind of "helps" he promises to give the research scientist on how to proceed with induction. All of these suggestions are already in evidence in his other writing.

We may find a clue among his 1613 remarks on the *Wisdom of the Ancients*. He depicts the "Sphinx, or Science," as a poser of riddles set for us to solve.

> Sphinx is said to propose various difficult questions and riddles to men, which she received from the Muses; and these questions, so long as they remain with the Muses, may very well be unaccompanied with severity, for while there is no other end of contemplation and inquiry but that of knowledge alone, the understanding is not oppressed, or driven to straits and difficulties, but expatiates and ranges at large, and even receives a degree of pleasure from doubt and variety; but after the Muses have given over their riddles to Sphinx, that is, to practice which urges and impels to action, choice, and determination, then it is that they become torturing, severe, and trying, and, unless solved and interpreted, strangely perplex and harass the human mind, rend it every way, and perfectly tear it to pieces. All the riddles of Sphinx, therefore, have two conditions annexed, viz., dilaceration to those who do not solve them, and empire to those that do. (Bacon 1884/1609, 411)

The clue here is not the passage's main point that some knowledge is urgent and practical. "For he who understands the thing proposed obtains his end, and every artificer rules over his work." (Bacon 1884/1609, 411). The clue in this passage is the reference to riddles associated with the Sphinx. We must pay attention to the riddle-like nature of the questions posed to us. When he says they tear the mind to pieces, he seems to echo the fear of the Pyrrhonian skeptic, who would abandon the riddle for a little peace of mind—*ataraxia*.

The main help that Francis Bacon provides to recognize the riddle is to ask us to record our natural histories in the forms of tables. He hoped to provide "helps" to the human understanding by this kind of tabulation. Other than a general description of these tables, there is not much more to go by.

A tabulation is a way of recording observations and experiments in some specific order. He tells us that any results we obtain should be formulated in tables of presence and exclusion and a table of degrees. These tables are a clue as to how natural history was collected. For if we are to collect facts randomly, there is no reason to expect that we could usefully tabulate them as "table of presence" and "tables of exclusion." When we first try to find a thing under some noted conditions and then look for an absence of the form under very similar conditions, we will have an instance of presence and another of exclusion. If we did this under many conditions, we would have several cases of presence and absence. If we have cross-examined nature thoroughly, perhaps we can have a table of both presence and absence of a nature and degrees of their manifestation of that nature. But if tabulation is a desired ordering, he does not tell us enough about what kind of ordering and how it will help us understand the true nature of a thing. After all, there are usually many more ways to order a set of elements than there are elements to be ordered.

We are, however, given a clue: we are to organize a list in such a way as to match the complementary list so that for each presence, there is a corresponding absence if one has been experimentally found. However, we are still stuck with nothing more than the list. There is some risk that the list will get us nowhere. At this stage, we are supposed to work out, somehow, what the nature of the thing is. We may even hope to obtain a better understanding of it to manipulate it to our benefit. Francis Bacon gives us an example of heat, in which he makes a small list of illustrative observations and experiments, puts them into tables, and asks us to consider the true form of heat on that basis. It is a singularly unhelpful example because, for one thing, it is unclear how his theory of heat is an induction from those tables. It has baffled learned commentators on Francis Bacon for nearly four hundred years. If Francis Bacon cannot state his method, his example is still more baffling.

Francis Bacon was such a brilliant writer that he hid his method in plain sight. While the image of the Sphinx-like verbal riddle is not quite the same as that of the labyrinth, it shares that character of a puzzle that Francis Bacon wants us to notice in the things to be understood. Speaking of how we must find out the latent or hidden processes and configurations or schematisms of the things we study, for example, he compares the process to a fire: "Therefore a separation and solution of bodies must be effected, not by fire indeed, but by reasoning and true induction, with experiments to aid; and by comparison with other bodies, and a reduction to which mix and meet in the compound. In a word we must pass from Vulcan to Minerva if we intend to bring to light the true textures and configurations of bodies on which all the occult, and as they are called, specific properties and virtues in things depend, and from which, too, the rule of every powerful alteration and transformation is derived." (Bacon II.7). The Vulcan heat of the experiments is produced by the skeptical inquirer, no doubt, but how do we use reasoning by "true induction" to obtain the wisdom of Minerva? This reasoning is not of the deductive kind found in the *Prior Analytics* of Aristotle, but more like the reasoning of a puzzle solver.

Francis Bacon was also known as a cryptographer, who invented a biliteral cipher, which he describes in the sixth book of the *Advancement of Learning*. His prowess is described quite independently of his new method as one mode of communication. The grammar of our language governs the normal mode, but when a secret message is sent, it may be governed by a cipher, such as the one he invented while he was in Paris as a young assistant to Queen Elizabeth's ambassador. In his cipher, two typefaces are used, differing so slightly that no one would ordinarily suspect it. Then, if these are alternated in a message, you will have a sequential arrangement of the two types. Suppose that one typeface is called 'a' and the other 'b.' Then we will get a combination of a's and b's as the hidden code. If five of these stand for a letter (there are enough combinations so that each letter has a unique combination of a's and b's), we can encode a message in the variable typescript at variance with the normal meaning of the message. Thus, he constructs a message that reads, in English, "Do not go until I come," but the varying typescript is encoded to say "F L Y."

Cryptanalysis, or the decipherment of texts, has also been argued to be the true method of induction, though with respect to the writings of Isaac Newton rather than the man he followed, Francis Bacon. Tyrone Lai (1989) has argued with merit that Isaac Newton's method of inductive inference was that of cryptanalysis. Francis Bacon was urging it. Later, Peter Pesic also applied this insight to Francis Bacon's thought (Pesic, 1999.)

He tells us that just as the mariners' compass needle helps with navigating better the oceans in which the sailor ventures, we need a compass to take us through the study of Universe, "...but before we can reach the remoter and more hidden part of nature, it is necessary that a more perfect use of the human mind and intellect be introduced." (Bacon 1857/1603, 33). What is this new compass that he calls the true induction that will help us decipher the hidden secrets of nature?

All this is strangely suggestive, but not quite enough to make his method of true induction plain. Francis Bacon seems to anticipate this when he says (*New Organon*, II.34) "Even to deliver and explain what I bring forward is no easy task, for things in themselves new will yet be apprehended with reference to what is old." It is not easy to guess his new clue when he does not set it out clearly and explicitly.

II

Philosophers faced with this mysterious impasse have made two plausible suggestions about the method proposed. One is that the Interpretation of Nature is an argument by the method of elimination—i.e., eliminative induction. Another is that his new method is a combinatorial composition of simple natures. Francis Bacon says things that superficially read in these two ways.

"But the induction that is to be available for the discovery and demonstration of the sciences and arts, must analyze nature by proper rejections and exclusions; and then after a sufficient number of negatives, come to a conclusion on the affirmative instances—which has not yet been done, save only by Plato, who does indeed employ this form of induction to a certain extent for the purpose of discussing definitions and ideas." (Bacon, I.105). The first part of this quotation suggests but does not say that by excluding all those impossibilities, we can affirm as true that hypothesis which is left. The method of proof by elimination, or eliminative induction as it has been called is a possible reading of the passage in isolation. However, the trouble with eliminative induction is that it is not an induction (*epagoge*). It is a form of deductive or analytic argument.

The logician H. W. B. Joseph (1916) proposed that Bacon's method eliminates every hypothesis to explain the facts until just one remains. This method of establishing the last hypothesis standing is famously also invoked by Arthur Conan Doyle as the method of Sherlock Holmes, though he does not call it the "Method of Eliminative Induction." As an argument, the form of eliminative induction can easily be expressed as a deductive argument. When it is valid, it is just a variant of the *modus ponens* argument, "If A is the case then B is the case. A is the case.

Therefore, B is the case." No one will likely mistake *modus pone*ns as an argument form that is anything but a deductively valid form. Nevertheless, Baconian induction is so murky that distinguished philosophers and logicians have at times allowed that the method of eliminative induction may have been what Francis Bacon had in mind as "helps" to the understanding in his account of induction from particulars.

Consider an argument of this form that we may call "MVP". Let A and B stand for statements. Then the argument is as follows.

1. Either A or B or both.
2. Not- A.
3. Therefore, B.

We will see that if 1 and 2 are true, then 3 must follow. That the inference scheme MVP is a variation on "*modus ponens*" is an elementary theorem in propositional logic, where "If A then B" (what is called the material conditional) and "Either not A or B or both" are considered equivalent. In an earlier development of elementary mathematical logic in the 1920s, this form of modus ponens, given as MVP, was used when logicians were leery of using the more common "if-then" form of a statement known as the material conditional (Hilbert and Ackerman 1950/1928). The method of Eliminative Induction allows more than two propositions that are alternates. Consider an argument beginning with several alternates in a statement assumed true. Let the argument eliminate all the alternates in it, except one. Then that one is true. The form of such an argument is still equivalent to the schema for *modus ponens*.

After all that Francis Bacon says about the futility of seeking a deductive account of first principles, he would be unlikely to fall back on yet another deductive argument as his grand new method—moreover, one that he claimed no one had ever tried before. We must concede that he could not have recommended this knowingly.

Did he not understand that the method of eliminative induction is actually a deductive strategy? If there are many alternative explanations, say ten, and we eliminate nine of them, then the tenth becomes true. Because there are often many possibilities, we may speculate that Francis Bacon got confused and thought he had a new method when he did not.

But this supposes that he did not grasp propositional logic, whether intuitively, as we all do, or as a theory. *Modus ponens* in particular and propositional logic more generally was worked out in ancient Greece by Diodorus Cronus, who taught both the Academic Arcesilaus and the founder of Stoicism, Zeno. We would need to show that Francis Bacon had not come across it.

Most commentators on his thought who have acceded to H. W. B. Joseph's analysis do so mainly because it has seemed that there is no better interpretation of Francis Bacon's method as he describes it here: "But the induction which is to be available for the discovery and demonstration of sciences and arts, must analyze nature by proper rejections and exclusions; and then, after a sufficient number of negatives, come to a conclusion on the affirmative instances…." However, we must read the quote in context. Here is what comes before and after the quote.

> …the induction which proceeds by simple enumeration is childish; its conclusion is precarious and exposed to peril from a contradictory instance; and it generally decides on too small a number of facts, and on those only which are at hand. But the induction which is to be available for the discovery and demonstration of sciences and arts, must analyze nature by proper rejections and exclusions; and then, after a sufficient number of negatives, come to a conclusion on the affirmative instances—which has not been done as yet, save only by Plato…. (Bacon, I.105)

While the method of eliminating all other possibilities is logically a feasible method, Plato cannot be construed as providing it. Plato's (or the Socratic) *elenchus*, or method of refutation, leads to perplexity, and recognition that we lack the knowledge we are seeking. Francis Bacon is quite aware of this aspect of the method.

> The school of Plato … introduced *Acatalepsia*, at first in jest and irony, and in disdain of the older sophists…. But the New Academy made a dogma of it and held it as tenet. And although theirs is a fairer seeming way than arbitrary decisions, since they by no means destroy all investigation, like Pyrrho and the Refrainers, but allow some of the things to be followed as probable, though of none to be maintained as true; yet still when the human mind has once despaired of finding the truth, its interest in all

> things gets fainter, and the result is that men turn aside to pleasant disputations and discourses and roam as it were from object to object. Rather than keep a course of severe inquisition. But … the human senses and understanding, weak as they are, are not to be deprived of their authority, but to be equipped with helps. (Bacon, I.67)

There is nothing in here that suggests that Francis Bacon's Plato practiced the method of eliminative induction. No doubt, his *elenchus* does exclude or eliminate. The list of possibilities considered must be complete for an argument from eliminative induction to be valid as an induction from particulars. We must know not only that "A, or B or C …" is true, but moreover that there are no possibilities other than A, B, C, that may be true. How can we extract this general claim of possibilities from the particulars we study? We see here that something is amiss because if we assume a completeness clause, it amounts to exactly what he asks us to avoid doing. A completeness clause anticipates nature. Moreover, if we rely on our imagination or past theories to generate all the alternatives, we can never learn anything new about the world.

> There is therefore much ground for hope that there are still laid up in the womb of nature many secrets of excellent use, having no affinity or parallelism that is now known, but lying entirely out of the beat of the imagination, which have not yet been found out…. [O]nly by the method by which we are now treating they can be speedily and suddenly and simultaneously presented and anticipated. (Bacon, I.109)

"And this very thing may be taken as an argument of hope, namely that there is a great mass of inventions still remaining which not only by means of operations are yet to be discovered, but also the transferring, comparing, and applying those already known, by the help of that learned experience of which I spoke, may be deduced and brought to light." (Bacon, I.110) The sheer novelty he anticipates from discovery rules out the possibility that the help to human understanding is eliminative induction. The method of eliminating all other alternatives works only if our mind supplies us with a limited number of cases, of which we must eliminate all but one.

The reasons why the method Francis Bacon is proposing is not the method of elimination is fivefold:

1. When valid, the eliminative method is a deductive argument, not an induction of any kind.
2. When valid, its premises must include a general statement (namely, that all the possible alternatives are those given of which all but one will be eliminated), which is an anticipation of nature. He rules that out.
3. If the general statement were incomplete, it does not go beyond the Socratic method of generating perplexity, or *akatalepis*. It is explicitly Francis Bacon's contribution, he says, to show how we are not left in perplexity. We need to have *eukatalepsis* or true grasp.
4. It is hard to see where we are to obtain the novelty that Francis Bacon promises us from his new method. If he believed that we cannot imagine the new reality for ourselves, then he cannot ask us to imagine it as a possibility before we can even begin.
5. There is a clear link that Francis Bacon sees between his new method and the inventions that we can expect to be able to use practically. This link is also inexplicable if all he proposed was the elimination of abstract possibilities that we had previously imagined.

We must conclude that elimination is a part of his method since he mentions it, but it is not the whole of it, as some commentators have hastily concluded.

III

There is ample evidence in Francis Bacon's writing to suggest that he believed in a combinatorial calculus of forms as the object of inductive study. "On a given body, to generate and surperinduce a new nature or new natures is the work and aim of human power. Of a given nature to discover the form, or true specific difference, or nature-engendering nature, or source of emanation, (for these are the terms that come nearest the description of the thing) is the work and aim of human knowledge."

With this beginning he sets out to disclose the inductive method in Book II, though he stops short of doing so in the incomplete *Novum Organum*.

His method needs interpretation and extrapolation. He seems to require that we first identify all the "simple natures" in the Universe. Every permissible combination will then give us an account of all the composite natures that can exist in the Universe. From this, one can derive a true description of the essential nature of all things. In Mary. Hesse's clear and well-argued interpretation of the method, Francis Bacon assumes that a finite number of simple natures need discovery from examining sufficiently many particulars. Accomplishing it would be the main task of true induction. (Hesse 1965).

Once these elements are at hand, the combinatorial possibilities would give us both insight into nature and power over it. Francis Bacon makes four assumptions, says Mary Hesse, and she is right: (1) that there are a finite number of simple natures; (2) that they can be fully enumerated in tables; (3) that their detection by the method of exclusion through observation and experiment is possible; (4) that the "form of a thing is in the nature of the thing itself," in Francis Bacon's words. In the case of heat, he gives us an illustration of how it is done. His conclusion that "heat is motion" shows that "not that Heat generates motion or that motion generates Heat … but that Heat itself, or the *quid ipsum* of Heat, is motion and nothing else." (Bacon, II.20). Thus "true specific difference" mentioned earlier would be how the definition of a nature is given in terms of a form. From this, It would seem that the proposed method of induction presupposes the existence and discoverability of simple natures, as Mary Hesse has observed. We need to work out simple natures and the forms that their true specific difference. Working them out is feasible only under the assumption of a finite, simple, and accessible set of simple natures that will in their combination, give us all the natural knowledge we seek. Is he suggesting a chemical cosmology? Mary Hesse was right—he was.

Chemical cosmology is very ancient in its conception. The nature of a thing dictates its typical motion, its natural motion or mode of change.[1] On Aristotle's account of it, four qualities were first among all that determined the nature of matter. These were two pairs of opposites, being hot and cold and wet and dry. That which is cold and wet is water, that cold and dry earth; warm and wet is air, and warm and dry is fire. A second list of qualities may include the pair "heavy and light." But these properties are dependent on the first, or so it was claimed. For the world is organized into the natural spheres as follows: the material of the earth is naturally at the centre of the cosmos, in the form of a sphere; the water enfolds it within a concentric sphere around it, the air, in a concentric sphere around that, and the fire around them all. The celestial spheres then go around them, with the sphere of the moon, the sun and planets and the fixed stars beyond that.

However, once we take Copernican astronomy seriously, this simple model breaks down, as Galileo Galilei saw. He showed that if the earth is just a planet moving around the sun, we must conclude that the clouds that follow it around are not light but heavy—after all, they follow the earth around. All bodies following the earth are heavy. They seem to gravitate towards the earth, irrespective of their analysis into elements. In modern philosophy, we are used to thinking of laws that govern motion itself. These laws are independent of the chemical nature of a moving thing. The movement itself is more fundamental, and the nature of things is secondary. Francis Bacon had no trust in this new mechanical conception of the Universe. He may have heard of it directly from his contemporary Galileo Galilei or indirectly through mutual acquaintances. But he makes no concession to its existence in his writing.

The cosmic distribution of the elements was an older description of matter. The nature of things was thought to generate the overall structure of the visible cosmos. Other kinds can then be explained secondarily. Heavy and light are to be understood as follows: a predominantly earthy piece will tend to go to its natural place and, therefore, will be felt heavy in air or water. In contrast, a bubble of air in water will tend to rise, as it is light therein. While the reader will be familiar with these accounts as medieval cosmology many times retold nowadays, its adaptation by Francis Bacon is still a bit of a surprise.

The ancient problem was this: Ptolemaic astronomy could accurately account for the positions and trajectories of celestial objects using specific mathematical techniques. Aristotle (following Plato's *Timaeus*) had given a widely accepted account of the motions of the planets in terms of their uniform, circular motions, which explains why the motions seem eternal.

But there was a clash between the two great subjects that could not be easily resolved. Computational astronomy, to be adequate, explained the retrograde motions of the planets by using a composition of deferents and epicycles, but to be quite exact, they also had to resort to a device called the equant. While the epicycle-deferent system conformed with the principle of uniform perfect circular motion, that of the equant was not. An equant seems to be moving uniformly to an observer sitting in some artificially defined place. The actual angular velocity would be disturbingly variable to a person going along an equant. Copernicus used the method of deferent and epicycle without using any equants when he suggested the sun-centered system of the planets.

Francis Bacon seems to have adopted a different account of the celestial motions than the Ptolemaic, in which the planets' motions were considered impeded rather than uniform to achieve the same results. Now, this method was not mathematically worked out well enough to have interested astronomers, but it would give us a physical theory of the motions that would be quite interesting for its time.

He also suggested that the world was composed of elements not divided as Aristotle thought, in terms of earth, water, fire, and air, but in terms of the alchemical trichotomy put forward by Paracelsus, of sulphur, mercury, and salt. If one pursues them one finds several appreciative things that Francis Bacon says about this analysis of elementary matter, as opposed to that of Aristotle. The cosmology and its details are partially drawn from Telesius and partially from Aristotle, but very eclectic, as Graham Rees analyzed it (Bacon 2000/1626.) Francis Bacon intended to produce this in the fifth part of his writings on the Great Instauration, he tells us, and this was just before the sixth and final part in which the true natural philosophy was to be propounded, though perhaps this would be completed long after he had died.

A very exotic speculative philosophy may have been intended for inclusion in the fifth part of his Great Instauration. Francis Bacon may have believed it to be true. But he was always careful not to suppose that his philosophy is the final truth discovered by his inductive method. In this respect, he is, like Plato, a skeptic about his own opinions. His speculative philosophy is also at a disadvantage by a criterion he emphasizes.

Whatever the method helps discover, it must pruduce *works*. His speculation delivers none. It is best to find this said in his own elegant writing.

> First, then, I must request that men not to suppose that after the fashion of ancient Greeks, and of certain moderns, as Telesius, Patricius, Severinus, I wish to found a new sect in philosophy. For this is not what I am about, nor do I think that it matters much to the fortunes of men what abstract notions one may entertain concerning nature and the principles of things. And no doubt many old theories of this kind can be revived and many new ones introduced, just as many theories of the heavens may be supposed that agree well enough with the phenomena and yet differ with each other. (Bacon, I.116)

There is no doubt in his mind that his speculative philosophy does not yield the works that he hopes the true philosophy will yield. "And I do not seek to found a school, so neither do I hold out offers or promises of particular works. It may be thought, indeed, that I who make so frequent mention of works and refer everything to that end, should produce some by way of earnest. But my course and method … is this—not to extract works from works or experiments from experiments (as an empiric), but from works and experiments to extract causes and axioms, and again from those causes and axioms new works and experiments, as a legitimate interpreter of nature." (Bacon, I.117)

Francis Bacon thus announces and denounces some substantive speculative opinions on the natural world. It is inevitable that when he propounds the application of his method in the future course of science, perhaps even after his death, his statements will assume some of his substantive conception of nature to state his views. How else could he write? But he tells us that this is not his aim, but this is him falling short of his purpose,

A striking feature of Francis Bacon's writing is that he does not confuse his method with his opinions. In this respect, he is a true heir to Plato.[2] Their epistemologies are different. Yet, he has as little regard for the attraction of philosophical sects as Plato had shown for the charm of sophistical persuasion.

One must conclude then that Francis Bacon thought his method was better than the expressions he was obliged to use. His words betrayed his opinions on cosmology and alchemy. He had to describe the world somehow in launching the Great Instauration. For this reason, the careful thoughts of Mary Hesse and the resourceful scholarship of Graham Rees still did not get to the heart of Francis Bacon's clever new idea for a method. They are right about his non-evident opinions. Their logical acumen and scholarly research are excellent. But even if they were right about his opinion on how the world seems to be structured, it should not matter to his theory of true induction.

For an inductive result to be true, it must be faithful to the measure of the world. When we follow an opinion and are convinced, we mistake our measure for what is real. Francis Baconls method of induction precluded him from advocating such an opinion. It is the method itself that must find the path to new knowledge. When we apply it, the new organon takes us to the promised land—a land where the new knowledge will also produce new *works*. And it is the land where the sufferings of the unfortunate are alleviated by the works created by the new knowledge. But in the meanwhile, Francis Bacon must describe his method as best as he can using misleading words at his disposal. Like all words, his words are full of Idols of the marketplace. He has no better words as yet to say his piece. The difficulty of prognostication is severe, which may be one reason why he did not get much beyond the second part of his six-part *Instauratio Magna*.

Notes

1. Harvey Wheeler stresses the importance of schematismus (which I call a model) and processus (typical motion) for Francis Bacon.
2. Samuel Taylor Coleridge called Francis Bacon the "British Plato." (Coleridge 1934/1818).

References

Bacon, Francis. (1603) 1857. Valerius Terminus of the Interpretation of Nature: With the Annotations of Hermes Stella. In Vol. 3 *of The Works of Francis Bacon*, edited by James Spedding, Robert Leslie Ellis, and Douglas Denon Heath. London: Longman & Co.

———. (1609) 1884. The Wisdom of the Ancients. In *Bacon's Essays and the Wisdom of the Ancients*, ed. Basil Montagu. Boston: Little, Brown, and Company.

———. (1626) 2000. A New Abecedarium of Nature. In *The Instauratio Magna: Last Writings*, edited by Graham Rees. Vol. XIII of *The Oxford Francis Bacon*. Oxford: Clarendon Press.

Coleridge, Samuel Taylor. (1818) 1934. General Introduction: Or Treatise on Method. *Encyclopaedia Metropolitana*. v.1 (1-43). ed. Edward Smedley, Hugh J. Rose, Henry J. Rose. London: Fellowes and Rivington.

Hesse, Mary B. 1965. Francis Bacon. In *A Critical History of Western Philosophy*, ed. Daniel John O'Connor. New York: Free Press.

Hilbert, David, and Ackerman, Wilhelm. (1928) 1950. *Principles of Mathematical Logic*. Tr. Lewis M. Hammond, George G. Leckie, F. Steinhardt. Ed. Robert E. Luce. New York, N.Y.: The Chelsea Publishing Company.

Joseph, H.W.B. 1916. *An Introduction to Logic*. 2nd ed. Oxford: Clarendon Press.

Lai, Tyrone. 1989. How We Make Discoveries. *Synthese* 79 (3): 361–392.

Wheeler, Harvey. 2001. The Semiosis of Francis Bacon's Scientific Empiricism. *Semiotica* (133): 45–67.

7

Evidence and Its Refutation

I

Francis Bacon tells us there are two ways to study nature: *anticipating* it or *interpreting* it. He said that every past school of thought has tried the first way. But he is the first to suggest the general method of *interpreting* nature. He allows that anticipations are far more powerful than interpretations for winning assent, i.e., as rhetoric, because though collected from a few familiar instances, they "straightway touch the understanding and fill the imagination." (*New Organon*, Bacon I.28). Francis Bacon thinks that we are naturally prone to anticipate nature. We need to restrain our understanding to get down to the more down-to-earth task of interpreting things as they are. In a much-quoted passage, he says, "The understanding must not therefore be supplied with wings, but rather hung with weights, to keep it from leaping and flying. Now this has never yet been done; when it is done, we may entertain better hopes of the sciences."

Therefore, his first and most general critique of all earlier philosophical schools is that they are anticipations of nature. They are opinions, or dogmas, as the skeptic may describe them; therefore, he would ask us to withhold assent.

J. Hattiangadi, *Francis Bacon's Skeptical Recipes for New Knowledge*,
https://doi.org/10.1007/978-3-031-52585-8_7

"In sciences founded on opinions and dogmas, the use of anticipations and logic is good; for in them the object is to command assent to the proposition, not to master the thing." (*New Organon*, Bacon I.29). His task is to advise us on how we may get hold of the thing that we wish to study, "master it," rather than merely be persuaded of something about it, which is pleasing to the imagination. This expression "take hold of the thing" is very much like the Stoic notion of "kataleptic" knowledge or the mental "grasp" of a thing perceived. It is precisely what the skeptics denied. But what Francis Bacon says about how we grasp the true nature of a thing is not by perceiving it in some particular way, for instance, innocently and without prejudice, but by conducting many observations and experiments.

The method that all philosophers have employed, until this newly proposed method of interpretation, he writes, is to proceed by anticipating nature. Opinions we loyally support can and will often face difficulties. But we can find, if we wish, that we do not need to abandon refuted general opinions. Our ingenuity knows no limit when finding ways to defend beleaguered opinions.

Instead, Francis Bacon proposes that we study nature by looking at a few things, whether well-known generalities or evident particulars. First, we must get to the lower-level principles, which he calls the "middle axioms." There are many of these, but as we advance up the pyramid, or the ladder of ascent, fewer and fewer principles will be admitted as ever higher-level axioms. After that, we proceed to higher stages. Before we reach the highest level of generality, at last, we must stop. We cannot try to know about faith, God's purpose, and how to live our lives. That would be committing Adam's sin. Short of theology, the method will take us in stages as high as we wish (Bacon I.19).

We can use deductive logic, of course. We can use the principles established at a higher level to determine the answer to lower queries, as in Aristotle's demonstrative model. The new method of interpretation of nature, Francis Bacon says, has as its main advantage that it uses induction to ascend to principles. Then, we can use syllogistic logic to work out its effects or consequences. It will give us fruitful experiments that we can use to improve the lot of all human beings. Since Aristotle developed the first account of the logical theory, the new method is Aristotelian in its descent. Only the ascent deviates from Aristotle's method in the *Organon*.

The difference resides in what Aristotle called *epagoge*. The new form of Induction is about establishing the first principles correctly. It may have been unfortunate that Francis Bacon used the word "induction" to describe his method, for the name has contributed to its misunderstanding. It is unlike *epagoge* and even less like induction and probability in the empiricist tradition after John Locke. But it is understandable why he used the word. He was challenging Aristotle's *epagoge*.

Plato described the method of hypothesis in the *Meno* as a method used by geometers. He repeats this description in the *Theaetetus*. A hypothesis is assumed to be clear and true and then used to shed light on its consequences. The "hypothetico-deductive" method combines it with the *elenchus*. Such a method, which philosophers of science have discussed, is also found in Pato's mature works. Karl Popper was an eloquent expositor of the virtues of this method. He tried to improve the method by suggesting that science consists of bold conjectures and their conscientiously solicited refutations. Karl Popper suggests that the method of conjectures and refutations, in his modified form, is the characteristic method of modern science. His remarks are quite the opposite of Francis Bacon's inductive description. However, where they both discuss the method of hypothesis, Karl Popper's remarks seem to be anticipated in Francis Bacon's description. We should study how Karl Popper understood refutations to understand Francis Bacon's account in Karl Popper's scheme. The difference between the two lies in this detail (Popper 2002/1963.)

Suppose the best modern science uses the method of anticipation, notwithstanding Francis Bacon's critique of it. In that case, we need to supplement it to make it work so that it yields good science. Most critics of the method of offering bold conjectures and refuting them in rapid succession offer us theories of induction or confirmation to supplement the method of hypothesis. Such supporting additions do not seem to help us understand the success of modern science, as Karl Popper energetically pointed out throughout his professional life.

Francis Bacon's objection to anticipating nature is that it is too easy to defend anticipations from their refutation. To this, Karl Popper adds that it is a virtue of a scientific theory that it is formulated to be easily refuted or falsified. Theories so formulated that they are not easily falsified are not good theories because we will not find out if they are mistaken. If the method of anticipation is the only method available to us, then the best that we can ever do, it seems, is to find out our errors as quickly as possible. We make our anticipations bold (or easily falsified) and formulate them in an easily falsifiable manner so that we bring them to bear honestly on the facts. Karl Popper invokes a moral or normative ethos of modern science as the chief reason for its comparative success. But refuting our great scientific theories may be harder than we first imagine, as he notes.

In the late nineteen hundred and sixties, Norwood Hansen, also subscribing to the hypothetico-deductive understanding of modern science, captured an insight succinctly by coining the phrase "theory-laden observations" (Hanson 1965/1958). Norwood Hanson's insight was that observations are theory-laden in that when we observe, we do not just observe what is there but filtered through our prior preconceptions. Observation reports are, therefore, riddled with hypothetical assumptions, even if we are not conscious of making them. Of course, this insight could also be argued as the unreliability of perception as a basis for knowledge. This unreliability was well known to skeptics, as witnessed by the Pyrrhonian arguments recorded by Sextus Empiricus among the ten Modes of Aenisedemus. Doubts about perceptual knowledge also form part of examining sense perception as knowledge in Plato's *Theaetetus*. Francis Bacon, too, suggests that the mind is like a false or distorting mirror (Bacon I.41). But Norwood Hanson's succinct phrase "theory-laden" compels attention.

Suppose we combine the hypothetico-deductive doctrine of science with the theory-laden or hypothetical character perceptual knowledge. In that case, we appear to be left in a sea of hypotheses. We have no facts beyond hypotheses to check them. Even if we wish to follow Karl Popper's suggestion to refute our ideas as quickly as possible towards a better understanding, how do we manage that? The refuting instances also seem to be hypothetical. Modern science seems hopelessly lost among its hypotheses. It is not easy to see how we can obtain any knowledge of reality if there is no empirical basis that is not merely a verbal convention of calling some assumptions "observations." We can pit one hypothesis against another, but there is no means for closing the debate except by a decision. How can we get closure that does not invoke a convention?

Some may have thought it reasonable to become skeptical about the claims of modern science. Many did. There were also many defenses of refutability in the face of theory-ladenness. There was a spirited discussion on this odd state of affairs in the second half of the twentieth century. Francis Bacon had a clever idea in the sixteenth to defuse this issue. He had a way around what has become a modern impasse, though he proposed it four hundred years ago.

When Francis Bacon wrote, modern science had started. Some may say it had not yet begun, and others that it was beginning. Skepticism about science was just as pernicious then as now. It was not easy to explain how science would gain knowledge of reality. It isn't easy after four hundred years, though we have modern science in our past. Yet we seem lost among our hypotheses, which fall short of knowledge. They seem to miss reality—not touching it anywhere.

Francis Bacon suggested a new technique for modern science, which was mainly a recommendation concerning the construction of Natural Histories. They formed the first part of his method of the Interpretation of Nature. Using a technique that may be dubbed called the "Method of Appearances and Refutations," it promised to overcome the theory-ladenness of observation more than three hundred and fifty years before its discovery. In one instance, Francis Bacon calls it putting Nature "on interrogatories" (Bacon 1857 v. 4). Questioning nature sounds odd. Still, the Baconian method was interpreted long after he died in terms of "queries" being put to nature. Isaac Newton appends a list of such Queries to his Optics as part of his examination of fundamental phenomena in experimental philosophy.

Francis Bacon's central idea is *to refute the appearances* or *the evidence*. These are not his phrases, and they sound odd. When we record the appearance of the nature of a thing, if we think that it conforms to its underlying reality, we call it our "evidence" of what is real. When we decide it is not like reality, we may call it "our mistake." Evidence and mistake are both categorized by the Pyrrhonians as "appearance," and we are asked to live by the appearances. In this manner, they say, we achieve peace of mind—*ataraxia*.

Francis Bacon's technique of cross-examining nature is to refute the appearances of a thing by contrary appearances of it and to note them all down in tables of conflicting evidence. Since we do not know which entries are true and which mistakes, we may describe them all as (conflicting) evidence until we solve the puzzle. We do not try to *avoid* errors in our evidence but to *gather and record it.* Then, we decipher it all at once for the true reality underlying the thing under study. However, if observations are theory-laden, we should not be surprised that appearances can be refuted, just as a theory can be denied. We will re-examine tainted evidence to see where it may mislead us.

Francis Bacon avoided writing much about skepticism, so his rare pronouncements are revealing. He also avoids talk about appearances in general. He was a great stylist of the language, however. He produced the same effect on his readers as the skeptics did by describing the task of science and its method paradoxically. Science, then and now, tries to describe abstractly some underlying reality of the world in which we live. It does what metaphysics did in his time. Francis Bacon recommends that we construct our metaphysics bit by bit: We ascend the ladder from the low base of the pyramid of knowledge. Writing to his contemporaries—all brought up studying scholastic philosophy as undergraduates—he used its language on how we scale the ladder from beneath.

At each stage or level, we describe the reality we apprehend in terms of the "natures" of things. Of course, one kind of thing has only one nature. It may be extracted from ordinary scholastic observations or the alchemists' esoteric experiments. When discussing dueling or differing natures of the same thing, he is alerting his readers that these are not true natures but apparent natures that conflict. Conflicting natures are not the true Ideas or Forms in God's mind, but the misleading *eidola* as they appear in human minds. Therefore, the method of experimental natural histories records the appearances or bits of evidence of reality *and their refutations.*

Appearances and Refutations do not suggest a plausible account of a viable method, like *Conjectures and Refutations* (Popper 2002b/1963). It is only a part of a new method, perhaps why it has been misunderstood. Refuting appearances or challenging the evidence seems either too defensive or too skeptical to count as a method of gaining knowledge. If we have conjectures, we may refute those by the appearance of things or the evidence. Challenging refutations seem the opposite of good sense and plain intellectual honesty. But whatever could we say about the value of refuting positive evidence? That they are theory-laden shows us why they can be refuted, but their actual refutation seems counterproductive. Having refuted the evidence, we seem to be left with nothing. To what end do we refute an appearance we take as evidence? If we succeed, we will no longer have those resources to refute our conjectures or hypotheses. Since confirmations are inconclusive for any general hypothesis science, we have effectively given up the only hope for grounding science in reality. Here is another reason why Francis Bacon's ideas are hard to grasp but also why, once learned, they are so enlightening.[1]

Appearances can be put on a scale of degrees, from being closest to perceptual experience at one end, like sensory evidence, to being far removed from perceptual appearances at the other end of the scale. A medical practitioner, like Sextus Empiricus, may notice that a patient has an unusually warm body and a flushed face. He may say that the patient appeared flushed and feverish, close to reporting a perceptual appearance. If recorded as they appear, symptoms of an illness stand in contrast to the diagnosis of an underlying condition of the patient, which would be a claim about reality. Sextus Empiricus may also prescribe ingesting a particular dried fruit that has been ground up with some herbs. He would prescribe this as a known remedy. It appears to work. Many medical practitioners may have treated their patients with the same symptoms in that way. If the patients find some relief, this fact qualifies as a seemingly successful treatment.

Of course, no one person has perceived all these events. To say that this compound of herbs or that tincture *seems* to have been effective in treating these symptoms or those in the past stands in contrast to diagnosing the actual illness of a patient. Today, we may perform tests, such as blood or urine tests, to look for an underlying cause. When we think that we understand the cause of an illness, we want to find a cure for it. We undertake actions to counteract the underlying condition, the etiology, which we would contrast with mere symptomatic relief. The etiology would be just what the empiric wants to refrain from asserting. The empiric prescribes by appearances alone, avoiding the headache of working out the etiology.

We recognize that the farther away from perception an appearance is, the easier it is to challenge its claim that it appears to be the case. The less it is like the evidence, the more like an opinion. Francis Bacon's idea, however, is not to challenge only those appearances that are easiest to challenge but also those closest to perception in our scale. In perspective, it is like Socrates asking his interlocutors to speak about what they trust to be true, what they sincerely believe, instead of just expressing an opinion.

His thought was novel, and he had to use many literary devices to show us how to do it. He gives us a description and an illustrative case study of the *Novum Organum* (in his analysis of the Form of Heat, starting at II.11). He calls them "Prerogative Instances." Later, several illustrative case studies in *Sylva Sylvarum* are also offered to show us what conflicting and surprising evidence is like. His examples illustrate the creation of histories of the use of discrepancies among appearances, i.e., in the evidence itself. These histories of differences in the perceived natures of things are without value if they are merely used to support a preconceived idea. He admits that we could dismiss some evidence and pick and choose what supports our preconceived ideas. We should, however, see how the avoidance of the "idols" of the mind is related to this exercise. We must also see how this effort differs from Aristotle's understanding of the nature of things by abstracting from their perceptions in the method sketched in the *Organon*.

There are always limitless conjectures that we could think up in time. Refuting each one serially, one at a time accomplishes very little by way of knowledge. By contrast, refuting an appearance, especially what we take to be a piece of solid evidence, is denying the truth of what we trust. Much like the Socratic *elenchus* in which a trusted ethical belief is undermined, here it is a trusted natural belief undermined. Such a refutation impacts our thinking significantly because we think the reality is such and so when we led to doubt it. The appearance of a thing commands assent to the seeming reality. It seems so even to after the skeptical challenge. But we challenge it with a discrepancy that is also compelling. That is the method.

It is not the end of the inductive method to be cussedly skeptical, but merely the means. "The doctrine of those who have denied that certainty could be attained at all has some agreement with my way of proceeding at the first setting out; but in the end infinitely separated and opposed. For the holders of that doctrine assert simply that nothing can be known, I also assert that not much can be known in nature by the way which is now in use. But then they go on to destroy the authority of the senses and understanding; whereas I propose to devise and supply helps for the same." (*New Organon*, Bacon I.37). Francis Bacon tells us why he is not a Pyrrhonian or an Academic skeptic, even though he seems to revel in engaging in the skeptic way.

II

There are two ways of reading Francis Bacon's advice on how to avoid the idols of the understanding. When he warns us about their dangers, what do we do about them? We can either be rid of them completely, once and for all. Or else we could keep them in mind so that they do not trip us up when it is inconvenient for us.

Pure minds can avoid these idols entirely, we may suppose. If only we could make our minds pure and free of Idols! We could then perceive things as they are. As John Locke later described it, a pure mind is like a blank slate, a *tabula rasa*. That is how Francis Bacon is also read by many today. It is engagingly summarized in Karl Popper's reading of this doctrine. Karl Popper has called this the theory that the "truth is manifest." Here is his succinct and well-expressed reading (references below as in Popper 2002b/1963):

> How can we prepare ourselves to read the book of nature properly or truly? Bacon's answer is: by purging our minds of all anticipations or guesses or prejudices (*Nov. Org.* i, 68, 69 end). There are various things to be done in order to purge our minds. We have to rid of all sorts of 'idols', or generally held false beliefs; for these distort our observations (*Nov. Org.* i, 97). But we have also, like Socrates, to look out for counter-instances by which to destroy our prejudices concerning the kind of thing whose true essence or nature we wish to ascertain. Like Socrates, we must, by purifying our intellects, prepare our souls to face the eternal light of essences or natures (cf. St

> Augustine, *Civ. Dei*, VIII, 3); our impure prejudices must be exorcized by the invocation of the counter instances. (*Nov. Org.* ii, 16ff)

> Only after our souls have been cleansed in this way may we begin the work of spelling out diligently the open book of Nature, the manifest truth.
>
> In view of all this I suggest that Baconian (and also Aristotelian) induction is the same, fundamentally, as Socratic *maieutic*; that is to say, the preparation of the mind by cleansing it of prejudices, in order to enable it to recognize the manifest truth, or to read the open book of Nature. (Popper 2002b/1963)

This reading of Bacon's intent in introducing the Idols of the understanding is not without apparent textual support. In Aphorism I.68, Francis Bacon writes, "So much for the several classes of the Idols and their equipage; all of which must be renounced and put away with a fixed and solemn determination, and the understanding thoroughly freed and cleansed; the entrance to the kingdom of man, founded on the sciences, being not much other than the entrance into the kingdom of heaven, whereinto none may enter except as a little child." Karl Popper also suggests a close connection between the Socratic and Baconian methods. But the relationship is not a truth seen in Neoplatonic light that is revealed to the pure of mind, as Karl Popper suggests. Francis Bacon was not following Augustine of Hippo in this manner. He follows a different side of Plato's philosophy. He was an Academic or a skeptic and not a Plotinian.

Francis Bacon did not think that the mind could be purified and freed from the idols once and for all. The clearest statement rejecting the thesis of pure perceptions of manifest reality is in the last paragraph of his fragmentary essay, *The Masculine Birth of Time*. "But my son, if I should ask you to grapple with the bewildering complexities of experimental science before your mind has been purged of its idols, beyond a peradventure you would promptly desert your leader. Nor, even if you wished to do so, could you rid yourself of idols by simply taking my advice without familiarizing yourself with nature. On waxen tablets you cannot write anything new until you rub out the old. With the mind it is not so; there you cannot rub out the old till you have written in the new. Nay, though you may possibly divest yourself of the *idols of the inn*, there would be every fear your falling victim to the *idols of the road*, unless you were prepared." (Bacon 1951/1601, 201)

In the *Novum Organum*, too, we find an echo of this opinion: the Idols of the Tribe may not be such as can be purged from the mind. "The Idols of the Tribe have their foundations in human nature itself, and in the tribe or race of men. For it is a false assumption that the sense of man is the measure of things. On the contrary, all perceptions, as well as of the sense as of the mind, are according to the measure of the individual and not according to the measure of the universe. And the human understanding is like a false mirror, which, receiving rays irregularly, distorts and discolors the nature of things by mingling its own nature with it." (, Bacon I.41)

This critique of human sense experience is like Plato's in *Theaetetus*. Every perspective of reality is rejected as prone to error. It is reminiscent of later skeptical philosophers of the Academy, and even the Pyrrhonians. Nothing could be more straightforward than this thought of Francis Bacon: the human mind, left unaided, cannot grasp a manifest truth in its perception of things, for it mingles its own nature (or the Idols of the mind) with things perceived. The mind, in its purest state, distorts reality. While the knowledge we seek must come from without, using perception, it must be extracted by a particular method. That method is not reliance on a pure perception, whatever that may be. The knowledge we can obtain from appearances must be extracted by a process Francis Bacon called *Induction*, or the *Interpretation of Nature*. There must be a place along this inductive path where the scientist is like an innocent child entering the kingdom of man. Francis Bacon says so, as quoted above. But it cannot be at the point at which perceptions are being recorded, as Karl Popper assumes. We must find its place later in the story.

We must remember that Francis Bacon was not writing his aphorisms steadily and in the order of presentation as some professional philosophers may write a text today. He was a busy man, involved in many law trials, state affairs, parliamentary affairs, and other affairs of other kinds. Consequently, he dictated short aphorisms when he found the time and assembled them later before sending his writing to print. Each aphorism is, therefore, like a piece of a jigsaw puzzle. We must place the pieces together in the right arrangement to make the best sense. Sometimes, one arrangement may seem evident until we realize that another is superior. Earlier commentary often got many things he said right, and it may even be full of insight, but it may not be exactly right because the aphorisms fit together into a better picture. There are many commentaries on the Idols, and almost everyone commenting on his work discusses them at length. But commentators seem to miss his reason for discussing them.

The Idols of the Cave are perhaps the most easily cleansed from the mind. To the extent that we have differences due to upbringing, it may be harder to eradicate what one anticipates. Nevertheless, in principle, they are the easiest of the four kinds of idols to remedy. But they have greater use than their distortion of reality (Bacon I.42 and 53–58).

The Idols of the Marketplace are much more pernicious, for our everyday language sometimes misleads us. Francis Bacon thinks that the philosophy of the Schoolmen is primarily what our common notions suggest to us. Consequently, an uncritical reliance on what our words seem to tell us traps us in scholastic philosophy. Francis Bacon's concern here is quite genuine. We can read it as Norwood Hanson's thesis of how theory-laden ordinary words describe our observations. New science will give us new propositions with new notions and concepts to describe its new evidence. This task is much more challenging to accomplish before the scientific results because we cannot know which vocabulary to adopt until after the science has advanced (Bacon I.43, 59, and 60).

Idols of the Theatre, such as the existing doctrines and their differences, may seem to be one target for "cleansing" from the mind, as Karl Popper interpreted Francis Bacon. But Francis Bacon wrote an entire book on the *Wisdom of the Ancients*, which respects past learning. He has this to say, too, about the wisdom of the past: "On one point not even a doubt ought to be entertained, namely whether I desire to pull down and destroy the philosophy and arts and sciences which are at present in use. So far from that, I am most glad to see them used, cultivated, and honored." (*New Organon* I.128). He does not complain about them that they are without value. It may sometimes seem that way from some of his isolated criticisms of each. He claims we cannot rely on them exclusively for future scientific advances. "Meanwhile, I give constant and distinct warning that by the methods now in use, neither can any great progress be made in the doctrines and contemplative parts of sciences nor can they be carried out to any magnitude of works." (Bacon I.128). If you do not destroy the old philosophies and science and arts, and yet if we cannot rely on them, then it seems that we must learn and doubt them simultaneously. He is recommending what may be called the "learned ignorance" of the skeptic to use Nicholas of Cusa's phrase.

Francis Bacon undoubtedly says that the Interpretation of Nature will yield knowledge of underlying processes ("latent processes" and "latent configurations" are his words, as in (Bacon II.1). But a pure perception in a clean mind cannot reveal this. No matter how clean the mind is, we cannot perceive realities hidden from our perceptions. It seemed to Peter Urbach that this theoretical or hidden and underlying reality could not be anything but conjectural (Urbach 1982). If they cannot be what the mind perceives after it is cleansed of prejudices, they must be supplied by the mind. They are conjectures. On this assumption, Peter Urbach offers many parallels between Karl Popper's writing and Francis Bacon's. He even suggests a striking methodological similarity with their dissatisfaction with the "pseudosciences," in Karl Popper's words. In Francis Bacon's thought, he interprets pseudosciences, on Peter Urbach's reading, as "anticipations of nature."

Peter Urbach's thesis may be well supported by appropriately arranging a few pieces of the jigsaw puzzle (the aphorisms). But if we look more closely, we find that anticipating nature is the same as the method of hypothesis when applied to the study of metaphysics. Francis Bacon complains that once a grand idea is arrived at in this method, it leads philosophers who are protagonists to defend and protect their favorite accounts from opposing arguments. They should be trying to find out their errors and correcting them. This advice has all the marks of Karl Popper's admonitions. Peter Urbach showed great insight in finding this affinity in Francis Bacon's scattered aphorisms.

Here is a difficulty. Francis Bacon's new method of the Interpretation of Nature will not be needed if we adopt a normative stand with the method of anticipating nature. We could adopt Karl Popper's normative stand that one ought to take refutations very seriously, even to the extent of courting them, rather than trying to explain them away. In this manner, we may at least find out when we are in error. We could also resolve to formulate our opinions boldly and falsifiably. By being ready to falsify them, we may avoid becoming too wedded to old beliefs. If all this were added to the method of anticipating nature, no new method would be needed.

But Karl Popper's opinion concerning method does not resonate with Francis Bacon's description of experimental science. He describes how we must spend a lot of time on "particulars" after doubting all opinions offered to explain them. "Now for grounds of experience—since to experience we must come—we have as yet had either none or very weak ones; no search has been made to collect a store of particular observations [that is] sufficient either in number, or in kind, or certainty, to inform the understanding, or in any way adequate." (Bacon I.118). If he were merely anticipating Karl Popper, he would have known the folly of taking in a lot of data. Only potentially refuting instances have any value in a hypothetico-deductive account of knowledge, as Popper argues forcefully. While Peter Urbach has many confirming instances of Francis Bacon's Popperian affinity, he cannot explain, on Popperian grounds, the negative instances found described in Francis Bacon's writings.

Neither Karl Popper nor Peter Urbach stressed an important feature of Francis Bacon's new technique of interpreting nature, though what they do find in it is full of insight: Francis Bacon often repeats that his new way will give us new works or technologically useful results. He describes this gain in power over nature as almost the same thing as his induction. And this also shows that something important is missing in these interpretations of Francis Bacon's writings. We could look at the history of science in retrospect and say that the technological or applied aspect of science is just serendipity. We could also suppose he is mistaken and merely bragging when he says he can give us better applications and superior technologies. But it would be foolish on our part to say both while ignoring how he thought his new method would give us power over nature.

There is a need to make more sense of what Francis Bacon is trying to say, and not just some of it, confirming our preconceptions. He believes that the mind is free from prejudices when disabused of *eidola* (Idols).

Then the mind naturally inclines us to learn something useful and enlightening at once. On the other hand, he does not seem to think that a cleansed mind can see nature transparently. There always more Idols. He promises "helps" for the mind to grasp reality and claims that his method provides "eukatalepsia," or a true grasp of things we seek to understand. The way of anticipations, or the method of hypothesis, only gives us verbal insight into things, he says, but what he promises and propounds is a method that "takes hold of the thing." Science as conjectures and refutations, or even as science as a gnostic perception of reality by a ritually cleansed mind, seems not to fit these oft-repeated remarks of Francis Bacon. The overlap with Karl Popper's thought is real but only partial.

III

Francis Bacon's method of observations and experiments or appearances and refutations stands in contrast to the method of Aristotle in the *Organon*. In that great work, Aristotle begins his *Categories* by classifying ten categories of existence based on what one would say exists. This exposition is followed (*On Interpretation*) by an analysis of the form of a proposition, which leads to the first formal account of logic ever produced (*Prior Analytics*), which in turn shows how we may argue validly from either true or false propositions, and that validity depends not on the subject matter under discussion, but on the form of the arguments used. It is after that point that we come upon *epagoge* or Induction.

At the end of this book and in the following (*Posterior Analytics*), Aristotle raises some crucial questions concerning knowledge. To understand something, we need to know why it must be true, he tells us. Understanding it is contrasted with an opinion that merely happens to be true. We recall that this was Plato's description of his uncompleted task in the *Meno*. An explanation of why something must be so can be shown logically or in other ways. Demonstration, using logically valid arguments, can give us certainty or necessity. We know that the conclusion must be true—but only provided that the argument's premises are also known to be demonstrated. This requirement leads to difficulty. How do we establish the premises to be demonstrably known? Aristotle provides us with an intriguing and influential argument.

There are some things that we know and others that we do not. Our task is to distinguish the cases. If demonstration were used to distinguish knowledge from true opinion—that is, if knowledge is proved but opinion is not—then we face a difficulty. Anything can be demonstrated by allowing circular ("reciprocal") arguments. This possibility contradicts our assumption that there are some things that we do not know. If we insist on demonstrations by ever new premises, then we reach the beginning of the proof only by an infinite regress, which is beyond us. In that case, nothing is known to be true. But, says Aristotle, if some things are known to be true but not everything is known, then we must avoid both viciously circular arguments as well as vicious infinite regress.

We can accomplish this feat if there are some principles whose truth is known of necessity in another way—that is to say, outside of syllogistic reasoning. These truths are the first principles of chains of demonstration. We arrive at them by a process of perception and abstraction. In this way, we get what can be called "definitions." They are statements of principle that describe the Ideas, essences, or natures inherent in things. To know the essence of a thing is to know that by virtue of which it is what it is. To grasp the nature or essence of a thing is, therefore, to grasp a kind of necessity, not something accidentally true of it. When we find that a person is human, for instance, we know that she must necessarily also be animal and rational. She must be so *necessarily*. This fact shows that we have found something that *must* be the case; therefore, it is known to us to be true. Essences or natures are based on perceptions but extracted by abstraction from memories of percepts by exercising the intellect. According to Aristotle, when we find the essence or nature of anything, we know why it must be what it is, which satisfies the requirement to constitute knowledge.

The method of appearance and refutation, evidence and discrepancy, or observation and experiment differs from the Peripatetic or Aristotelian way. We first work out what appears to be the nature of a thing under certain normal conditions. Let us call this an "observation." Then seek an experiment. The same thing is perceived as exhibiting a different nature in an experiment. We observe one thing in some ascertained conditions. The discrepancy is generated by observing a different nature in conditions close to those in which the first observation was made. We find a different nature in kind or degree from the first instance. This new way of proceeding, undermining the evidence, and questioning what has been observed is the prescribed Baconian method of experimentation. We note how similar it is to a Socratic refutation of an interlocutor's belief about a virtue. It is a cross-examination of the researcher to whom the evidence of nature seemed evident.

If any observation of a thing under study has a particular essential nature, the experiment under closely related conditions refutes it. The experiment shows that the thing evidently has a different nature even though the conditions are barely altered. But nothing can have more than one nature in the doctrine of Forms and Ideas. The method of appearance and refutation, or observations and experiments, is recorded in a natural history of a strange kind. Unlike a scholastic natural history of observations of specimens, an experimental natural history is an assault on sane common sense.

IV

The ten arguments or modes of Aenesidemus that the Pyrrhonian skeptics compiled are reflected in the method of Appearances and their Refutations. The arguments of Aenesidemus are also reflected in Francis Bacon's thoughts concerning the idols of the mind. But they are reflected in different ways. We can find a clue to Francis Bacon's thinking in his different uses of skeptical techniques.

The ten arguments of Aenesidemus were divided by Sextus Empiricus into two kinds: some were relativities (subjectivities, we could call them) and those that were discrepancies in things themselves (objectivities). Those arguments based on relativity are adopted by Francis Bacon and expanded to warn us against the pitfalls of the idols. The idols of the tribe and the cave are summaries of the arguments against dogmas that

Aenesidemus lists. The fourth class, the idols of the theatre, are the dogmas against which the Pyrrhonian skeptic marshals these ten arguments. Francis Bacon lists the dogmas and their influence upon us as one class of idols. There is nothing new here, merely a difference in presentation and emphasis. Sextus Empiricus tells us how to try and avoid the idols of the theatre without classifying them, whereas Francis Bacon classifies them roughly and asks us to seek ways to challenge them.

In the third class of idols, the idols of the marketplace, Francis Bacon goes beyond the ten arguments from Aenesidemus. That our ordinary notions themselves may mislead us is a startling thought and does add to our stock of skeptical concerns, though Sextus Empiricus would have approved. In opposing the old dogmas and the anticipations of nature, Francis Bacon and the Pyrrhonians are united in opposing the same schools of thought.

The difference occurs in using the discrepancies in Francis Bacon's method. When two appearances of a thing are different, Sextus Empiricus tells us that the skeptic withholds assent about the real. He asks us to live by appearances without resolving the discrepancies. If a thing seems to have a particular nature in one condition but another when perceived in a second condition, the skeptical response is to refrain from assenting to the thing's true nature. Instead, we live by the appearances. We note, as skeptics, that under condition A, it appears thus, but under condition B, it appears so. It does not matter that it *seems* thus and so, and that nothing can really be thus and so at once. There is nothing wrong with *seeming* incoherent. We refuse to assent to any claim of reality.

This difference is the clue to Francis Bacon's method of true induction. He savors these discrepancies and asks us to seek out more of them. We refute the appearances repeatedly. He thinks we can use its details to grasp its true nature when we find enough impossibilities about that thing. Then we can break through the appearances that deceive us to a reality formerly hidden beyond them. How the particulars noted down in natural histories like this can be used to winkle out the true natures of things is Francis Bacon's first step in the "method of ascent" from particulars to the middle axioms or principles. The step is then repeated as we ascend higher.

The basic idea can be summarized as follows, though not altogether in his own words. When our perceptions yield our knowledge of the natures of things, the Idols of the mind are mixed in with the idea of the thing itself. The nature we perceive is not its true nature but one intermingled with our nature to see what our mind compels us to see. (Here, again, is the theory-laden character of all observations in an earlier guise.) We can therefore find other perceptions nearby (or very similar) in which the thing studied appears to have a different nature in very similar circumstances.

Having found this, we must not take the final step, the stance of the Pyrrhonian skeptic, and merely assent to what *seems* to be the case. It is true that whatever seems to be the case to me or here or now need not be consistent with what appears to you there or then. Idols are many and diverse, and each is incomplete in explaining the thing studied. But reality, as opposed to Idols, is always self-consistent and a comprehensive account of appearances. That is Francis Bacon's clue to the nature of induction.

We take a path, he suggests, opposed to the one taken by the skeptics after following the skeptics thus far. His diagnosis is, in short, that skeptics gave up too soon. They conclude that reality is unreachable. The conclusion was premature. If we find errors, we need to record the mistakes. We keep recording them until we see something, at last, that is deeper but free of the recorded errors. How we find that elusive something is another step that must be taken—and an interesting topic.

Francis Bacon does not use the words "appearances" or "evidence" and their "refutations," though we can use them to interpret his writing. More common in his time were phrases like "observation," "report," and "experiment." Whenever an observation is reported, it is a statement of an appearance in the language of Pyrrhonian skepticism. We can use that partial overlap in meaning to interpret the results of any observations in Pyrrhonian terms. This practice can show us the importance of his thinking of the Modes of Aenisedemus or Plato's criticisms of sense experience as knowledge in the *Theaetetus*. So, we could replace "appearances" and "their refutations" with the more common expressions: observations and experiments. But this still leaves the word "experiments" to understand.

In Francis Bacon's time, the meaning of "experiment" was not that of a refutation of observation. This interpretation of experiments was his distinctive philosophical contribution. Experimentation was carried out primarily by alchemists, who sought to understand the esoteric and hidden nature of substances by these means. The guild of alchemists conducted experiments to reveal true and mysterious natures. Members showed experiments of this kind to other full guild members—usually only to them. An experiment shown to some initiates is a revelation of the true nature of something hidden in the appearances—and certainly also hidden from outsiders. The public would have to pay for the tradesperson's knowledge. As a trade secret, an experiment was distinct from any observation report that someone in the general public could also easily make. Experiments were opposed to the common understanding of things. Francis Bacon exploited this opposition.

Francis Bacon's invented his theory of induction for the reform of chemistry. Its greatest and most influential success came later in Isaac Newton's optics and physical astronomy. However, the method was a chemical method in its first conception. In chemistry, the opposition of experiment and observation manifests itself in the opposition between alchemical and Aristotelian natures. Francis Bacon's thought was to make this opposition a technique for cross-examining nature. Experimenting was not for establishing exotic truths, as in Alchemy. Experiments would question observed conclusions (scholastic natures) but without being conclusive about reality.

Later in the seventeenth century, this new technique came to be called the "experimental philosophy." But experiments in experimental philosophy were no longer what alchemists had intended. Experimentation was no longer a technique for drawing deep and exotic inferences. It was not a technique to be practiced in isolation from other observations and experiments. Philosophers began using it primarily to introduce a skeptical attitude at the observational level, interacting with things. Of course, the older association of experiments with extraordinary chemical discoveries did not entirely disappear. Most commentators on Francis Bacon who are amused at his outdated interest in the Renaissance art of alchemy will be surprised to note it is not he but they who have an alchemical and Hermetic conception of experimentation.

But how does the new technique give us knowledge of reality, and how does it yield power over nature? Those are topics that we must explore and seek to understand.

Note

1. The difficulties of interpreting what Francis Bacon and later experimental philsophers (Robert Boyle, Robert Hooke, Isaac Newton) meant by the word "experiments" have been highlighted in many publications by Peter Anstey, for instance in his (2014). The solution to his keenly described problem is that experiments must be understood as part of the cross-examination of nature that Francis Bacon describes elsewhere as part of the task of science.

References

Anstey, Peter. 2014. Philosophy of Experiment in Early Modern England: The Case of Bacon, Boyle and Hooke. *Early Science and Medicine* 19 (2): 103–132.

Bacon, Francis. (1603) 1857. Valerius Terminus of the Interpretation of Nature: With the Annotations of Hermes Stella. In Vol. 3 *of The Works of Francis Bacon*, edited by James Spedding, Robert Leslie Ellis, and Douglas Denon Heath. London: Longman & Co.

———. (1601) 1951. The Masculine Birth of Time. In *'Temporis Partus Masculus: An Untranslated Writing of Francis Bacon'*, translated by Benjamin Farrington. *Centaurus* 1 (3): 193–205.

Hanson, Norwood Russell. (1958) 1965. *Patterns of Discovery: An Inquiry into the Conceptual Foundations of Science*. Cambridge: Cambridge University Press.

Popper, Karl. (1963) 2002b. *Conjectures and Refutations: The Growth of Scientific Knowledge*. London: Routledge.

Urbach, Peter. 1982. Francis Bacon as a Precursor to Popper. *British Journal for the Philosophy of Science* 33 (2): 113–32.

8

Francis Bacon's *Elenchus*

When an abstract belief becomes orthodox in any community, it must subjugate philosophy, particularly skeptical philosophy. Nonconforming secular thought is a threat to it that supports dangerous heresies. A heretic, however, cannot hope to persuade faithful members to change their orthodox opinions on heretical grounds. The faithful are taught to guard against that possibility. A religious orthodoxy usually has well-developed ways to counter such a political threat. Christian orthodoxy, we must remember, was more than a thousand years old. No matter how rational, or reasonable, or factual, a nonconforming opinion is it cannot be persuasive enough to topple an authoritative orthodoxy that has withstood the test of time. A rational, reasonable, and factual objection to an orthodoxy only challenges the faithful to test their loyalty—their faith. The faithful must find it in their hearts to overcome the doubt and thus reaffirm their belief. If anything, a test of reason only strengthens orthodoxy. By overcoming test after test, our faith is burgeoning.

It is different when religious orthodoxy begins to loosen its hold. It happened in the eighteenth century in France. French philosophers admired earlier British success in secular science. Orthodoxy had been weakened and contributed to the Enlightenment in France. Pursuing free

J. Hattiangadi, *Francis Bacon's Skeptical Recipes for New Knowledge*,
https://doi.org/10.1007/978-3-031-52585-8_8

science was possible in the secular Republic that they adopted. But England in the seventeenth century had no recent example before it. It is almost impossible to imagine how secular thought could have freed itself from religious orthodoxy if its success was merely a wish. Nor was it a wish to escape when secular, skeptical thought became socially acceptable. There was an intermediate moment. It happened because local secularism made sense within Christianity in England—within its orthodoxy.

Francis Bacon's rescue of secular science and philosophy in Christian England on Christian grounds is not a paradox. It does not make science any less secular, nor the Church of England any less Christian. He argued that he was restoring our dominion over nature after we had been thrown out of Eden. This story has a faultless theological beginning. Furthermore, he argued that we should pursue secular science because the duty of charity, arising out of the Christian love of humanity, requires it of us. Universal love or charity has a special place in the gospel. It forms an appropriate motivation for modern science on Christian grounds. But such an argument could only be made acceptable to the devout flock of a Church if they could see how such a new kind of science would benefit humanity.

I

Plato's philosophy was a secular philosophy. He may have invented theology as it came to be adopted in Islam and Christianity, but he was pagan, he was kafir, he was goy. The Byzantine Emperor Justinian was also not Christian. But he led a Christian empire. When he banished the last iteration of Plato's Academy from his empire, he did so because it constantly challenged the official faith of his dominion. He did what the head of an empire could do to further its interests (Heather 2018). Of course, as Plato practiced it, philosophy challenged every opinion, including his own. When freely practiced, Socratic questioning is always skeptical of all opinions and every faith. It remains obsessively secular. Questions like that are problems for any proclaimed orthodoxy.

As it developed later, theoretical science became a specialized part—the reliable knowledge—of secular philosophy. It does not customarily clash with religious thought. In diverse cultures and civilizations, science (or philosophy) and religion do not clash, or rarely do so. Philosophy, which asks general questions of importance to human beings and speculates about likely answers, is found in all cultures and all religious groups. There is no particular tension between science and religion as such. Exceptions occur when any belief becomes orthodox and is defended as overriding any possible knowledge.

Two religions stood out for having ongoing difficulties with secular thought, especially skeptical philosophy. The gospel is not responsible for its use in the intolerance of philosophy—at least not in its original interpretation. Christianity, at first, was like any other religion. It became an ideological emblem of the Roman Empire when Constantine legalized it and appointed Christians to lead his government. Soon after, it became the official imperial religion of Rome. A particular reading of the holy books became a mandatory creed. The ideological emblem that took over in the name of Christianity after that could no longer tolerate free thought. Free thought only encourages heretics. Philosophy became dangerous because it could question orthodoxy, and orthodoxy bound the Roman empire together when it had multiple centers and constantly threatened to fall apart.

In a face-to-face, simple, communal religion, the common sort of religion found among humans, members of a religious community can identify each other by their common customs, language, and a network of kinship bonds. We classify such religions as tribal and unorganized. The members of such groups share ritual practices and the telling of historical and mythic tales within the community. Whether a member believes the tribal myths is hardly relevant to membership. They are already part of the group. What is appropriate to membership in a religious community is how people behave in critical situations, which allows them to continue as members. People who act in strange ways are left out of communal events to different degrees. But their belief is not what matters. If they repudiate the group, they may disqualify their membership in extreme cases.

When Christianity became the state religion of the Roman Empire at its alternate seat in Byzantium, however, it had a significant burden to carry. It had to become the identifiable abstract emblem of a far-flung empire at war over many centuries with the equally powerful Persian (Sassanid) Empire to the east (Crawford 2014). Both realms were exhausted from war. Both had trouble keeping away the smaller forces of local hordes battling them while distracted by the larger conflict.

Constantine's new orthodox religion of the Roman Empire could not use kinship, common rituals, customs, or even a shared language to bind the far-flung empire. Tribal religions are too local for empires. The regions within the Roman or Byzantine Empire were very different. What remained to bind Christians together was an abstract faith. It must have been a discovery that an abstract belief can be made an emblem of unification. It can then be used and reused to reinterpret pagan customs along Christian lines. This ploy makes conversion less onerous for the converted. They do not need to overturn their lives as much. Thus, Christianity spread comparatively easily in the old Roman empire and beyond. People could accept an abstract belief to enjoy the protection of Rome while keeping their language and customs more or less steady.

The main casualty of this abstract form of empire is philosophy. For any religion to become an ideological emblem and a political force, it must first become orthodox—that is to say, the empire must adopt a right-thinking religion. Those who are to be unified must hold the religious doctrine precisely as prescribed for an empire to unify around it. Soon after Constantine made the Christian religion unofficially an orthodoxy, he called the Council of Nicaea. A new Christian orthodoxy was debated, determined, and later officially enforced. Bishops—those who favored orthodoxybecame thought police, to use a modern phrase, casting nonconforming bishops from the Roman Empire.

Philosophy had become a dangerous pastime to the unity and resolve of the Christian Empire. No longer could one freely question the meaning of the gospel and examine speculations as early Christians had done. Skeptical thought is especially dangerous to the unity of an ideological empire. We are not surprised when we learn that the Christian Emperor Theop-hilus destroyed the Great Library of Alexandria. On the other hand, if the good Emperor did not succeed entirely, then perhaps Muslims did it when they won back Alexandria. There is a story of a Khalif being persuaded that if the books in the library did not reflect the teachings in the Quran, they should be destroyed, but if they did, they were not needed. Nothing surprises us about anti-intellectualism, though how and by whom the great library was eventually destroyed is unknown. The Muslims under the Umayyads soon adopted the same technique as Christians did to try to hold their empire together. Keeping together a vast and growing empire is a solemn business.

Islam developed as an orthodoxy in formerly Persian Asia, just as Christianity had in Rome. A powerful family (the Umayya) who opposed the prophet Muhammad during his lifetime formed the fifth Khalifate and soon embarked on reconquering the old lands of the Persian empire in the name of Islam. Islamic Khalifates were then and shortly after able to contain the spread of Christianity in the east, and to a lesser extent in Europe, by getting Christians to pay an additional tax in the territories they controlled. The Mediterranean was orthodox.

Getting their orthodox ideology right and enforcing intolerance of philosophical variation in faith became a recurring preoccupation of ideological Islam. Secular thought was also a casualty in Islamic Persia as it had been in Christian Rome. The Church Militant and Jihadi Islam were soon fighting again, like the Romans and the Persians of old. The political entities changed, but the wars remained. But now they were now engaged in a holy war for their true religions.

Each religion carefully nurtured science and philosophy, too. The nurture of philosophy emanating from Persia began in Abbasid Islam and carried over to medieval Christian communities in the West. Defenders of faith cite this nurturing as counterevidence to their intolerance of philosophy. Orthodoxy was not quite so anti-intellectual, we are told. The intolerance of ideas in orthodox religions must be tempered by this nurturing that happened, too. However, if we reflect upon it, we see that their philosophies provide even more evidence of their intolerance.

Each religion opposed secular thought or free philosophy, which they suppressed. But they carefully nourished and encouraged a *submissive* philosophy that they could use to bolster their orthodoxy. We see the rise of two paradoxical new medieval philosophies: Islamic Philosophy and Christian Philosophy. The defenders of orthodoxy carefully monitored submissive thought to keep it in check. The Dark Ages were kept dark by the thought police. The comparative spread of ideas in medieval thought was not because of but despite the orthodoxies of their time and place.

II

We must marvel at the emergence of entirely secular natural philosophy or science in England in the seventeenth century. Francis Bacon proposed such a secular natural philosophy, which is relentlessly skeptical. It is as skeptical and secular as Plato's philosophy, which was an active fallibilism and had been banished from the Roman empire. Neo-Platonism survived and even flourished as a part of Christian orthodoxy because of an endorsement by Augustine, Bishop of Hippo. But Augustine preached to the converted. They had already abandoned skepticism. Plato was safe once he had his fangs removed.

How did Francis Bacon manage to introduce a skeptical method into Christian England? How is it that the Royal Society not only encouraged but protected the pursuit of philosophy free from religious concerns? All the evidence suggests that the Fellows of the Royal Society of London were faithful Christians. They had diverse convictions within that category. One of its founders and supporters, Robert Boyle, was a lay preacher.

The Royal Society of London was a rare thing. It had the protection of the state (of Charles II) to pursue natural philosophy wherever it led. There was no requirement for Fellows to observe any religious orthodoxy. Neither the countries with allegiance to the Church in Rome nor the Protestants, who believed in knowledge from the bible alone, *sola scriptura*, could keep their theology from interfering in natural philosophy. But Francis Bacon suggested differently—and the Royal Society accomplished it in his honor.

Francis Bacon devoted a lot of thought to religion. His faith did not allow him to subordinate Christianity to secular or skeptical philosophical thought. Nevertheless, his new inductive method allowed great freedom to the skeptical form of philosophy he adapted from Plato, which has come down to us as modern science. Francis Bacon's main concern was keeping his faith safe from the relentless skepticism he let loose in his natural philosophy or science. He wanted to separate science and religion in a certain way based on his faith, which allowed them to coexist.

He developed the Socratic *elenchus*, primarily a method of refutation, into his inductive method. Socrates had reserved its use to investigate moral questions. Plato's Socrates would ask someone to state their opinion concerning a virtue. He would then refute the assertion using what the interlocutor readily admits to knowing. It seems improbable that a Socratic *elenchus*, which helps us find errors in the opinions of others, could generate universal principles, whether of virtue or natural philosophy. Francis Bacon's new idea was about how to learn affirmatively from an *elenchus*.

Induction is about discovering principles, whose collection Francis Bacon likens to a pyramid. We begin our induction at the bottom of the pyramid, studying only modestly limited experience classes, one at a time. We extract the middle or lower axioms and gradually extract higher-order axioms or principles. We proceed without omitting intermediate steps. He said we scale the ladder of the intellect, climbing upward on the pyramid of science, by using the same method of a "true induction" applicable at the very first step. Each step is like a cross-examination of nature. Every new level reached is liable to be revised as we ascend higher.

It appears strange and unlikely that Baconian induction, extracting principles from experience, is like a Socratic *elenchus*. Yet, Francis Bacon asserts it, as we see in the numbered quotations below, which are taken from his writings.

1. "... But a really useful induction for the discovery and demonstration of the arts and sciences should separate nature by proper rejections and exclusions, and then conclude for the affirmative, after collecting a sufficient number of negatives. Now, this has not been done, or even attempted, except perhaps by Plato, who certainly uses this form of induction in some measure, to sift definitions and ideas." (Bacon, I.105)
2. "... In other words, I mean (according to the practice in civil cases) in this great Plea or Suit granted by the divine favour and providence (whereby the human race seeks to recover its right over nature), to examine nature herself and the arts upon interrogatories." (Bacon 1858/1620, 263)

His curious reference to cross-examining nature herself is his unusual way of using the Socratic *elenchus*. We are left to wonder in what language or manner Francis Bacon expects inanimate nature to respond to our queries or cross-examination. Even more intriguing is how the method of cross-examination can lead us to conclude on affirmatives after a series of exclusions and rejections without making gratuitous assumptions.

Francis Bacon fashioned his inductive *elenchus* out of a curious admixture of the thoughts of Plato and Aristotle. He accepts Aristotle's account of logic in *Prior Analytics* as a valuable instrument for drawing valid conclusions from premises. Logical deduction is used to apply established principles high on the pyramid of knowledge to appearances lower down on the ladder. He also concurs with Aristotle on the need for a different kind of demonstration that is not deductive to take us from particular appearances up to the general principles themselves. Deduction alone is insufficient to establish the principles of science. Induction must be invoked.

His language betrays some Aristotelian or scholastic influence when he writes on induction, epistemology, and metaphysics. Aristotle's account of a non-syllogistic form of demonstration (*epagoge*) consists of a perceptual grasp of the Idea or Form of a thing. Francis Bacon also allows for this as part of his new method. Following Aristotle, for the most part, he wants to find the essence or the "nature" of the thing. The nature of a thing is an attribute of the thing under study. Aristotle did not adopt Plato's understanding of Ideas or Forms as independent entities, separate from the things themselves, nor did Francis Bacon. Individuals capable of change are referenced as the primary entities, or primary substances, by Aristotle, and Francis Bacon's language assumes as much. Attributes have a secondary or dependent existence on primary individuals. A thing's Form, Idea, or "nature" is an unchanging feature attributable to a changing individual.

Following Plato and Parmenides, however, Aristotle insisted that we cannot understand change because it is indeterminate. The essential nature of anything (any individual) has a changeless potential that actualizes in time in its characteristic or natural motions. Francis Bacon concurs with Aristotle that the nature of anything being studied would have a characteristic motion or process, which would be its natural law of motion. Like Aristotle, he thinks of the reality of a thing as a fixed nature that it possesses. He also followed Aristotle in thinking that the nature of an individual is grasped by the mind and not perceived by the sensory apparatus. Though we access the nature of a thing primarily by perceiving it, we grasp that nature intellectually by abstracting it in our language.

Francis Bacon was an Aristotelian about general knowledge, but only up to a point. He allowed us to use a method of abstraction to find the principles or axioms governing a thing. He agreed that we can abstract their forms from things we perceive. But he rejected abstracting as a method that was too naïve if used by itself for a true science—i.e., one that is faithful to reality. We need something more to get at the true forms of things. Aristotle's method consisted of mentally stripping off, one by one, all its contingent attributes from the kind of thing we are studying. Having stripped away the accidental qualities, we are eventually left with the bare attribute of its essence. The thing we are studying will then have that nature by virtue of which it is what it is. Other inessential attributes are accidents of the item, of which Francis Bacon has little to add to what Aristotle had to say.

Francis Bacon admits that we could use this method of abstraction by the mental subtraction of superfluous attributes to formulate an account of the nature of whatever we are studying. However, he rejects it as the sole method that is satisfactory for induction. The abstractions we thus obtain are merely the common notions in use, expressed in the concepts of our everyday language. He found this kind of induction inadequate. He objects to the method of abstraction on two grounds.

(a) He objects that it is prone to error because it is based only on a few positive instances. The concepts of our language are abstract. Because of the poverty of abstraction, our everyday ideas mislead us even before we begin speaking of the natures underlying them.

3. "The syllogism consists of propositions, propositions consist of words, words are symbols of notions. Therefore if the notions themselves (which is the root of the matter) are confused and over-hastily abstracted from the facts, there can be no firmness in the superstructure. Our only hope therefore lies in a true induction." (Bacon, I.14)

This objection is rooted in his understanding of Plato's thought.

(b) He also complains that Aristotle's principles, in common with all earlier philosophies before Francis Bacon's time, have little practical application. By way of contrast, medicine and alchemy are useful, though in a limited way, and not as systematically as he would want. This complaint shows the influence of his Anglican religion.

4. "Of all signs there is none more certain or more noble than that taken from fruits. For fruits and works are as it were sponsors and sureties for the truth of philosophies. Now, from all these systems of the Greeks, and their ramifications through particular sciences, there can hardly after the lapse of so many years be adduced a single experiment which tends to relieve and benefit the condition of man… the experimental part of medicine was first discovered, and that afterwards men philosophized about it, and hunted for and assigned causes; and not by an inverse process that philosophy and the knowledge of causes led to the discovery and development of the experimental part…." (Bacon, I.73)

One of his objections is founded on religion. Francis Bacon's reference in quote 2 above to the human race seeking to "regain its rights over nature" is a clear reference to the biblical story of the fall. He hopes to help humans recover from our loss. This theme permeates his writing over decades. This recurring motif in his literary work is philosophically vital because it encapsulates his response to skepticism. He fashions a partisan, religious canopy under which his thoroughly secular conception of science can flourish. That is how he introduces secular thought into a devoutly Christian England.

III

After graduating from Cambridge University, he lived in France for a couple of years at the French consulate of the court of Queen Elizabeth. He was an impressionable young man of sixteen when he went to France. The Pyrrhonian Crisis raged in France then and continued for decades after (Popkin 1960.) He must have been a keen observer of it. He had read the first essays by the renowned skeptic Michel de Montaigne. Later, he adopted the essay form in English to become known as the first English essayist (Bacon 1985/1597). He was acquainted with the skeptical problem of the criterion of knowledge from reading Michel de Montaigne.

Richard Popkin (1960) has given an excellent account of the widely discussed question in the sixteenth and early seventeenth century concerning the criterion of religious knowledge, also called the "rule of faith." Skeptical arguments were first adduced in a partisan way to challenge one or two forms of the Christian faith (the Lutheran or Calvinist form) by another group (the Roman Catholics). That the bible can be interpreted correctly by anyone with a clear conscience seemed philosophically naïve to critics. What criterion would you decide on the correct interpretation if there was a disagreement? That the bible is its own criterion of correct interpretation seemed to be an elementary error of circular thinking. To Catholics, the sacred tradition of the church seemed to be the correct answer. But skepticism is sharp and double-edged when used like a sword to cut down dogma. What makes the sacred tradition sacred? Can we make sense of it without the gospel? Though it may be a wider circle, we must still invoke the bible to justify itself. All Christian philosophers have to take skepticism seriously as a challenge.

Such arguments lead to a "Pyrrhonian crisis." We find an interesting resolution to the crisis in fideism, as Richard Popkin explains it. Fideists accept the doctrine that the arguments of the Pyrrhonian skeptic against all knowledge claims beyond what appears or seems to be the case. A fideist endorses the Pyrrhonian recommendation to "live by the appearances." Living by appearances, we assent to what appears to be true. Then, we act on that basis. The skeptic withholds assent from any opinion concerning any non-evident reality. Living by the communal norms of ethics is a part of Pyrrhonism. We cannot endorse reforms to a custom because we would have to claim to know better: "What do I know?" Michel de Montaigne developed the most influential fideistic defense of the Christian faith in his erudite and brilliant essays. He continued to practice Catholicism as a skeptic. The sacred tradition represented, after all, the most widely practiced religious custom in Europe.

Francis Bacon dealt very differently with the Pyrrhonian crisis. For religious reasons, he suggests that we disallow the mixing of knowledge of the divine with the human. Once we have disallowed any skeptical examination of religious questions on religious grounds, we are free to adopt skepticism as a technique in philosophy and science on the same religious grounds. He does exactly that in his *Proficience and Advancement of Learning, Divine and Human* (1605). In a short, unpublished, and incomplete manuscript of 1603, Francis Bacon provides a pithy account of his biblical story of the fall.

5. "Man... when he was tempted before he fell, had offered unto him this suggestion, *that he should be like unto God.*" (Bacon 1857/1603)

Adam and Eve succumbed to the temptation of hubris, and after eviction from Paradise, lost all knowledge. But God, taking mercy on us, sends a representative to tell us his will, contained in the general instruction to humans:

6. "*Love your enemies; Be you like unto your heavenly father, that suffereth his rain to fall both upon the just and the unjust,* doth well declare, that we can in that point commit no excess." (Bacon 1857/1603)

He interprets the message from God (here capitalized by him, but italicized above) as a duty of charity to do whatever we can to help others, without limitation, except for restriction in the lesson of the fall itself.

7. "*All knowledge is to be limited by religion and to be referred to use and action.*" (Bacon 1857/1603)

To seek to know divine things is forbidden to us. It led to Adam and Eve's banishment from the Garden of Eden. But, he says, this does not preclude inquiry into natural knowledge. We may do so within the bounds prescribed by religion. His religion was Christianity. He was a member of the Church of England. Christians must pursue knowledge as a duty of charity, but not for its own sake. We must seek knowledge directed towards practical benefit to humans. It must be born out of love rather than pursued for the pure understanding of the mind. Only in such a way can we avoid Adam's sin of hubris.

He suggests that the Greek philosophies, admirable in many ways, were defective. They were conceived without the Christian duty of charity in mind. Consequently, they failed to be productive or beneficial to humans. Alchemists conducted experiments that brought some benefits to humanity. But they were haphazard in their search for knowledge. Unlike philosophers, the tradespeople valued their expertise for profit, not charity. Their search stopped when they found a piece of practical knowledge that they could exploit for gain. Both plans are defective. He saw the Christian task as finding luciferous or enlightening knowledge for more fructiferous or beneficial experiments to support Christian charity.

Francis Bacon saw differences between interpreting the Book of Nature and the Book of God. He interprets the Book of God as a plain account of how God wills us to live. After Adam's fall, God sent a special emissary to tell us His Will, i.e., how to live a good life. If any doubts remain about God's Will, King James I and his Bishops can settle the issue. For this reason, the pyramid of knowledge has a small pyramid at the apex marked off and reserved for the Church of England. However, the Book of Nature is not a plain statement of God's will. It is still a bounty of his provenance.

One of Francis Bacon's signal achievements was arguing for a secular philosophy as a Christian duty. It allowed secular philosophy to become unthreatening to religion. It allowed Christians to support secular science. There is still another hint in the bible, a very slight one, on how our charitable deeds may be useful to humans. Francis Bacon regularly refers to the Book of Job, saying for example,

8. "And if the book of Job be turned over, it will be found to have much aspersion of natural philosophy. Nay… Salomon the king affirmeth directly that the glory of God *is to conceal a thing and the glory of the king is to find it out*, as if according to the innocent play of children the divine Majesty took delight to hide his works, to the end to have them found out…." (Bacon 1857/1603, 220)

We can find out about nature by seeking what is hidden. But what is hidden in nature cannot lead to hubris, because we will not allow it to do so. His manner of separating the divine from the human is one of the main points of the first book he published on *The Proficience and*

Advancement of Learning, Divine and Human. Its central thesis is that when philosophers are rightly instructed, divine knowledge need not expect interference from human natural philosophy.

His religious thought also influenced his evaluation of past knowledge claims. He does not entirely dismiss them. Past thought has defects, which can be severe. When pointing them out, Francis Bacon sounds like a skeptic. But he insisted from the start that the wisdom of the past is not entirely without merit. He was opposed to the skeptic's renunciation of all philosophical knowledge. We can use the old thoughts only after we have taken note of their defects. However, he provided an optimistic assessment of their usefulness.

Fideism and Baconian theories of science are similar in this: both seek to separate skeptical philosophy from religious opinion. They differ in the manner of their separation. Consequently, they differ in the resulting scope of science. Under fideism, natural philosophy is confined or closely quarantined by skeptical doubt, perhaps undermining it altogether. Religion is allowed free rein to roam because of our faith in it. It can roam everywhere because it has suppressed those skeptical parts of philosophy that threaten all opinion. Baconian science is different. Science, taking up skeptical methods, is free to roam everywhere except in the narrowly confined region of divinity. Religion is closely quarantined to studying the meaning of biblical texts that teach us how to live our lives. Salvation is protected from skeptical doubt on religious grounds. Philosophers must avoid infringing on biblical interpretation. The reasons are derived from religion. Francis Bacon gives broad scope to philosophy and science because the recommendation to be charitable is boundless. We may love other humans without any limit. We can, therefore, pursue natural knowledge without restraint, provided we do so to alleviate the suffering of others.

IV

Aristotle's basis for understanding first principles relied on an intellectual grasp of perceived essences. Induction of first principles uses abstraction to find the nature of things. Francis Bacon dismisses this reliance as prone to error. To understand his objection's full force, we must study Plato's thoughts on knowledge, which influenced Francis Bacon. In quotation 3 above, just before he suggests that his method of induction is Socratic in form, he says this about Aristotle's method:

9. "...For the induction which proceeds by simple enumeration is childish; its conclusions are precarious, and exposed to peril from a contradictory instance; and it generally decides on too small a number of facts, and on those only which are at hand." (Bacon, I.105)

One of the delights of reading Plato's dialogues is that when he is not gleefully demolishing the claims of others, he takes the time to criticize his own. He is just as severe on himself. Plato's remarkable criticism of his own doctrine of recollection seems to have profoundly affected Francis Bacon.

The doctrine of recollection is introduced in the *Meno*. Meno asks Socrates why anyone would seek knowledge. He would not seek it if he already knew what he was seeking, for there would be no need. But if he did not already know what he was seeking, he still would not seek it, since he would not know it even if he were to find it. There are only two possibilities; in each case, one would not seek knowledge. Plato suggests a possible way out of this dilemma. If the soul already possesses the knowledge being sought but cannot recollect it, then gaining knowledge may be just the recollection of that knowledge. We would recognize what we seek when we find it by recollection. In a sense, we already possess it, enough to recognize it. In another sense of possessing the knowledge, we do not have it because we need to know it more readily. Plato opined that perhaps we forget what already resides in the soul because of the trauma we suffer at birth. This possibility suggests that the recollected knowledge may be innate. That we recognize something as knowledge is the criterion of knowledge.

The dialogue *Meno* displays some characteristic doctrines of Plato—the doctrine of knowing as recollection, the doctrine of innate knowledge, the doctrine of souls existing outside the typical life-cycle of a person, etc. However, the *Meno* does not include one philosophy that we associate most with Plato, namely that the knowledge we recollect is of

Ideas or Forms. They are introduced later as a hypothesis in the *Phaedo* in which the doctrine of recollection and the theory of Ideas are closely connected. In the *Republic*, these accounts of knowledge become an essential part of Plato's theoretical apparatus for studying justice in the city and the soul.

Many arguments in later dialogues question and test Plato's cherished earlier doctrines. Among his later self-critical dialogues, we must also count the *Theaetetus*, in which Socrates addresses the question, "What is knowledge?" He discusses several theories of knowledge but does not once mention the Ideas or Forms in that dialogue. He ends the dialogue after considering all the theories of knowledge and concludes that we do not know what knowledge is. However, as in the *Meno* , so in the *Theaetetus* Plato considers the doctrine of knowledge as recollection.

Imagine that bits of knowledge we possess in our minds are like birds flying about freely within an aviary we own. We possess these birds, since they are in our aviary. If each bird represents a piece of knowledge, then we have them all. Perhaps these birds represent that which we possess innate to us. In another sense of the word, we do not possess them at all since they are not in hand, but flying about in our aviary. We do not know, in that second sense of the word. But is innate knowledge a criterion of the manifest kind of knowledge? Plato's answer is, surprisingly, no it is not.

Suppose we catch a bird. Have we not learned a bit of knowledge and so learned something by having it in our hands? The *Theaetetus* does not mention the earlier doctrine of recollection. We can recognize that the aviary is his old model for it. At this point, Theaetetus asks Socrates a profound question: What if some birds do not accurately represent reality? What if we do not know by holding them in our hands? Could some of them represent ignorance, or nescience? Socrates commends the question and Theaetetus for asking such a good question, but he does not give an answer that meets the need. He has none. The dialogue continues with a consideration of knowledge as true belief and of knowledge of true belief with an account or a logos. When these are both shown to be unsatisfactory, Socrates and Theaetetus conclude that we do not know what knowledge is.

This critique of recollection contained within the dialogue entails that we cannot rely on an innate Idea or principle just because it is there. A recollection may seem to illuminate what we want to know, but it may only serve to distort it if it is a bird of ignorance. There is no mark or criterion of a true light of the mind that could distinguish it from a false light. In Francis Bacon's terminology, translated from Plato, we cannot know whether the bird in the aviary is an Idea, a part of God's plan for things in the world, or a distortion of it in our mind, created by an *eidolon*, or an Idol.

10. ".... Be it known then how vast a difference there is…between the Idols of the human mind and the Ideas of the divine. The former are nothing more than arbitrary abstractions; the latter are the creator's own stamp upon creation, impressed and defined in matter by true and exquisite lines." (Bacon, I.124)

If our innate ideas can be false or distorted, then perhaps we cannot know anything at all as it really is. Whether we mentally grasp the nature of things through perception, as Aristotle believed, or whether we grasp things by the inner light of reason, as Plato thought earlier in his philosophical career, we face the same vexing difficulty. We cannot know reality as it is with our corrupt minds. We cannot trust our minds, most basic intuitions, plainest perceptions, and convictions. Francis Bacon seems to have extracted from the *Theaetetus* that, unless aided by a new method, there is a desperate difficulty in seeking knowledge. The mind cannot know without assistance. We may express this negatively by asserting that *there is no safe haven basis for knowledge*. We may note that this desperate condition also follows from the biblical story of the fall. When Adam and his progeny were cast off, they were without language, knowledge, or even a moral compass to live out their mortal lives. We were in the desperate situation that Plato described in the *Theaetetus*. Out of abject despair is born hope.

An unfinished Baconian manuscript of 1601 is entitled "The Masculine Birth of Time." The enigmatic title refers, I suggest, to Socrates. Socrates says in the *Theaetetus* that whereas his mother was a skilled midwife who assisted women when giving birth to their children, Socrates, though barren of ideas himself, had a special ability to help men give birth to new ideas. This "masculine birth" was accomplished precisely by Socrates' use of the *elenchus*. Francis Bacon, reading Plato many centuries later, wants to credit Socrates with assisting in the birth of his new idea—an ingenious idea—of how the *elenchus* can help us obtain true inductive knowledge of the universal principles of things. This may explain why the manuscript is entitled "The Masculine Birth of Time" The phrase "of time" in the title seems to be an attempt at modesty. He seems to suggest that he is not taking credit for the discovery, but merely noting that its time had arrived, and he happened to have learned it. In the title of his unfinished manuscript "The Masculine Birth of Time," he is paying homage to Socrates.

The strange title of his manuscript reveals a strong gender bias. There is no reason why the birth of ideas has to be, or is, masculine. Plato's Socrates, however, speaks of inspiring men; following his lead, Francis Bacon seems to have chosen to interpret his ingenious new idea as a "masculine birth" long after the death of Socrates. This slender clue tells us how to interpret his *New Organon* in crucial matters. For he thinks, like Plato, that the birds in the aviary may be Forms or Ideas, or they may be Eidola, and we cannot tell which, even if, or especially if, both are innate.

In several of Francis Bacon's writings, we find Idols mentioned. In the *New Organon*, he lists a four-fold classification of Idols. The Idols of the Tribe are those we possess by the very virtue of being human. Try as we may, we cannot remove them from our minds, for they form part of our nature. Then there are the Idols of the Cave, which beset us when we are unwilling to venture into the light and are content to be guided by subjective images of things, each to their own. The Idols of the marketplace are all those Idols that arise from what we may also call social factors. He particularly mentions the language in use. Words that signify distorted concepts of things are Idols that can mislead us in our search for true Ideas (I,e, faithful to reality.) Yet, we must rely on these words to commence any science. He had not heard of the social construction of reality, but it would have been high on this list of Idols of the marketplace. He also mentions the Idols of the Theatre or the lecture theatre. The opinions and arguments of famous philosophers and past scientists are propounded there. Philosophers can persuade us with well-argued opinions before we inquire about the truth. They can distort our grasp of the genuine Idea.

Every learned commentary on Francis Bacon's method notes these Idols. Many paraphrase his descriptions well. But they miss his point—i.e., why Francis Bacon mentions them at all and does so in such detail. A common misunderstanding of the Idols is to suppose that he wants us to purge our minds of them, once and for all before the work of Baconian science can begin. Once cleansed, a mind is as a newly waxed tablet, or it is as a clean slate, on which true characters may be written by perceived nature, as John Locke could have said; or that the human mind, once cleansed of Idols, is a true mirror of nature, as Richard Rorty could have expressed it. This is what Francis Bacon had to say, however:

11. "On waxen tablets you cannot write anything new until you rub out the old. With the mind it is not so; there you cannot rub out the old till you have written in the new. Nay, though you might possibly divest yourself of the *idols of the inn*, there would be every fear of your falling victim to the *idols of the road*, unless you were prepared." (Bacon 1951/1601, 201)
12. "For it is a false assertion that the sense of man is the measure of things. On the contrary, all perceptions as well of the sense as of the mind are according to the measure of the individual and not according to the measure of the universe. And the human understanding is like a false mirror, which, receiving rays irregularly, distorts and discolours the nature of things by mingling its own nature with it." (Bacon, I.41)

In quotation 11, we find two Idols not mentioned elsewhere in his writing: the Idols of the Inn and the Idols of the Road. In the *Advancement of Learning*, he only discusses three of the four idols of the *New Organon*. The fourth is mentioned but not taken up. He does not seem systematic about the classification of the Idols. Francis Bacon was not trying to find a true classification of the Idols, as some commentators have supposed. His message is different. The idols are numerous, ubiquitous in our minds, and belong to very different kinds. Therefore, their errors cannot be avoided, which is the first point of the Idols.

The unavoidability of the Idols plays an integral part in his ingenious idea of induction by cross-examination. From the unavoidability of the Idols and their distortions of our minds, we can find no haven - no sure foundation - to ground knowledge. If we must have a secure learning source, then there is no hope. That desperation must be admitted. But out of desperation arises new hope. The diversity of the Idols, the second point of mentioning them in detail, and the independent sources of their origin, is a source of renewed hope after we have almost conceded to Pyrrhonian skepticism. The Idols' diversity, abundance, and mutual independence are central to his method of "true" induction.

V

Francis Bacon's ingenious idea is that we learn from our errors if we are patient enough to wallow in them. He recommends what we may call the method of "appearances and refutations" or of evidence and its refutation. His method suggests observation, experiment, conjectures, and refutations rather than trial and error. He insists that we should not try to avoid errors defensively or to explain them away when we find them. He asks us to record as many errors as possible that are honestly made in a limited study of particulars. He also asks to write down the conditions in which the errors are made.

A Baconian natural history records not just established facts, reputable facts, or unassailable facts but also unusual facts, surprising facts, and even fantastic facts. Of course, we must ascertain that they are facts and under what conditions they are found faithful to reality. The kind of natural history Francis Bacon wants us to record includes discrepant natures. They *appear* to be real, though they cannot be because of their mutual incoherence. When we observe a thing, we are asked to record the observation of the evident nature and the conditions under which it is observed. His experimental method recommends that we seek out and record another condition close to the observation where the same thing exhibits a different nature. The unusual nature of the same thing in a similar documented condition constitutes a prescribed experiment.

Nothing can both have and not have its nature. A successful experiment generates a discrepancy in the apparent nature of the thing under study. In effect, an experiment refutes an evident fact about nature. It is a paradoxical metaphysical micro result.

Francis Bacon's ingenious suggestion begins with this experiment. Faced with a perceptual discrepancy, we are tempted to act. We want to resolve the difficulty. But acting too soon is a fallacy. We commit a new fallacy that we may call the "Baconian fallacy" in his honor, for he discovered the inductive fallacy of *jumping to conclusions* or *rushing to judgment.*

In a deductively valid argument, it matters little whether we jump headlong to the conclusion or slide into it with stealth. In a valid argument, we can rush to conclude and not invalidate it. Logicians call validity "monotonic"—adding more premises to a valid argument makes no difference to its validity. It is not so with "true" induction. There is a persistent tradition among scholars of Francis Bacon's writings to read his induction as "eliminative." It was first proposed by H.W.B. Joseph (1916), and found ready acceptance among logically minded philosophers and historians of thought.[1] However the induction proposed by Francis Bacon is not monotonic. Logicians will note that this is fatal for understanding Baconian induction as eliminative. We constantly revise our axioms as we ascend the ladder of knowledge. The task of expanding our natural history does not end with one induction. Science, or knowledge, is not achieved in a day.

Interpreting nature like Solomon begins with patience. Our solution to a puzzle depends on how many components we have asembled. We do not try to understand the nature of the thing at once, nor do we give up the search prematurely. The first would be dogmatism, which commits the fallacy of rushing to judgment; the second is skepticism, or rushing to avoid judgment; each is premature, and each fallacious.

13. "....The understanding must not, therefore be supplied with wings, but rather hung with weights, to keep it from leaping and flying. Now this has never yet been done; when it is done, we may entertain better hopes of the sciences." (Bacon, I.104)

If we rush to judgment, we will adopt a philosophical dogma. We can then defend by explaining away any future discrepancy. We can protect our doctrine with ingenious tricks. An opinion can always be protected, "come what may," according to the motto of a latter-day holist (Quine 1951). But this will not allow our mind to "take the measure of" reality. Instead, we are measuring the universe by a dogma born of the distortion caused by the idols of our minds.

14. "The human understanding when it has once adopted an opinion (either as being the received opinion or as being agreeable to itself) draws all things else to support and agree with it. And though there be a greater number and weight of instances to be found on the other side, yet these it either neglects and despises, or else by some distinction sets aside and rejects…." (Bacon, I.46)

To avoid dogma, we may be tempted to avoid commitment, like a Pyrrhonian skeptic. When we refrain from dogma, in order not to rush to judgment, we may give up hope and think to ourselves that we should withhold assent from everything. Such restraint may seem to be the only way to avoid dogma. We must show restraint, but we must also patiently restrain our restraint.

15. "It will also be thought that by forbidding men to pronounce and to set down principles as established until they have duly arrived through the intermediate steps at the highest generalities, I maintain a sort of suspension of the judgment… *Acatalepsia*—a denial of the capacity of the mind to comprehend truth. But in reality that which I meditate and propound is not *Acatalepsia* but *Eucatalepsia*". (Bacon, I.126)

Adopting a skeptical response is premature because it is a rush to *acatalepsia*, not unlike a rush to judgment. It is the same fallacy, though its object is withholding assent, rather than adopting an opinion. We will have missed the true Idea hiding in the experimental natural history. We can only find the thing's nature when we are willing to look for it. The

Pyrrhonian skeptic gives up too quickly, lured by the promise of possible *ataraxia*.

Francis Bacon does not tell us the measure of an adequate natural history. He advocates patience. But how long a wait is long enough? There is no calculation possible until we know what is hidden. His *Salomon's House* implies a surprising communal solution to this question. In that scientific utopia, the study of natural philosophy is organized communally, as Dana Jalobeanu (2015) has stressed in her sensitive reading of Francis Bacon's thoughts on natural histories. There are the producers of natural histories, organized in a community that keeps on cross examining nature. They do not stop for anyone. Then there are the theoreticians who interpret the record, in a different unit of Salomon's house. Presumably, the theoreticians can tell when a natural history is ripe enough to interpret its nature. When sufficiently ripe, the history or account will support a unique solution. If premature, many interpretations of the natural record will be equally possible. Natural history is ripe when it cannot support a skeptical decision to withhold assent between two or more possible interpretations. We can be Pyrrhonians no longer.

We need to organize the appearances of a thing *in all its distortions* to use them for induction. We must put them into tables of the presence of a given nature, tables of their absence in close proximity to their presence, and tables of its different degrees when they exhibit them. This sort of tabulation exhibits the discrepancies in the appearance of a thing's nature in full light. Tabulating the presence and absence of a nature in proximity amplifies the discrepancies to our mind. When we have a great many errors recorded, we will have created a riddle to solve or a labyrinth to navigate. We must find the true Idea and our way out of the maze, emerging without the distortions created by the idols of our wretched minds. Now we need the wisdom of Solomon.

Several metaphors describe how the last part of an induction is conducted. As we have seen, Francis Bacon uses the metaphor of God playing hide and seek with Solomon. In *The Wisdom of the Ancients* (1884/1609), he compares science to a sphinx that asks us a riddle. We must decipher an encrypted message when answering a sphinx-like verbal riddle (Pesic 2000). He says elsewhere that we must find our way out of our labyrinth. More recently, other writers have also spoken of this aspect of science. Albert Einstein has suggested that scientific discovery is like a detective using clues to solve a mystery (Einstein and Infeld 1938). Karl Popper likens it to problem-solving (Popper 2002b/1963), and Thomas Kuhn suggests that normal science is like puzzle-solving, perhaps a bit like putting together a jigsaw puzzle (Kuhn 1962). Tyrone Lai has written that the method of science adopted by Isaac Newton is a kind of crypt-analysis (Lai 1989). Peter Pesic suggests Francis Bacon also had crypt-analysis in mind (Pesic, 2000)

The metaphor of a key that fits the lock is also not so far-fetched. If the Book of Nature is like a coded message, then learning how it is encrypted, to decipher it, is to interpret nature. It may aptly be described as finding a key to a lock. We are urged to fashion a lock so only a unique key will fit. When we have used that key to unlock the treasure chest, we have interpreted nature. We note that this is not an anticipation of the mind; for the key has to fit all the locks that emerge in the assembled discrepancies of natural history (in a sequel to this book, it is likened to a passkey). Finding such a solution was the point of our extended "cross-examination of nature." We have not been content to record the testimony of our observations, like the scholastic, but experimented to discover their hidden discrepancies and, therefore, the puzzle. The tables help our minds find the right solution.

Francis Bacon did not finish his plan for writing the Great Instauration, so we do not have his final account of the last part of the method. But we can extract the rest of it from what he has said. His new form of puzzle-solving depends on three features of errors if they have been patiently recorded in Baconian natural history:

A. A distortion caused by an Idol is not merely an error; it is also an appearance of reality (albeit a distorted appearance.) It constitutes evidence. The method of appearances and refutations differs from conjectures and refutations. A refuted conjecture is mainly useless to us. A refuted appearance remains a source of knowledge of reality, even if we know it to be a distortion. Each distortion is also, at the same time, a facet of the reality we seek.
B. In a recommended natural history, many possible kinds of distortions of the nature of a thing could appear. Each has an independent source. A second distortion can refute an appearance distorted by a first. Each

observed nature of a thing pretends to be the whole of its nature. A thing can only have one nature. When a different nature in proximal conditions counters a thing's observed nature, each observation refutes the other. Every distortion is caught out in this way by a refutation in an extensive natural history. The Idols' great diversity and independent origin allow us to discern distortions of nature. The abundant diversity of the Idols is, therefore, of the greatest importance to his method. Francis Bacon might tell us that God has granted a bounty of errors. Now it is up to us to use this gift well.

C. No distortion of reality can exhibit the entire nature of the thing studied. Since the evidence is distorted, it exhibits only a facet of its true nature. Only the Idea itself will be a consistent account of all a given bit of reality. If a schematism faithfully captures its nature, it will describe it completely. Distorted, it is partial. No matter how many facets of a thing we adduce, its true nature, should we find it, cannot be contradicted by them. It is up to us to find the model that fits all the confusing evidence.

Given these three features, we can understand how the method works:

16. "On a given body, to generate and superinduce a new nature or new natures is the work and aim of human power. Of a given nature to discover the form or the true specific difference, or nature-engendering nature, or source of emanation… is the work and aim of human knowledge." (Bacon, II.1)

If we are careful in our observations and experiments, wallow in their errors, and note errors with the conditions of their occurrence, then the only consistent interpretation or model of the appearances will be the true nature itself. Every distortion or partial facet of the thing will be refuted by another distortion of appearance—because neither can be the whole of its nature. That is the solution of the puzzle, the way out of the labyrinth, the code deciphered. It is also what Francis Bacon means by finding the "nature-engendering nature" or "true specific difference" or "source of emanation" that we seek.

Experimental results look different before and after the solution. They seem to change after they are explained. When multiplying discrepancies, we face a puzzle. Like a Socratic interlocutor, we are paralyzed as if stung by a stingray. Once we have solved the mystery, everything makes good sense. Every phenomenon of a thing is seen as a facet of its true nature. The phenomena are normalized and compatible. We can identify each phenomenon by its place in the new schematism. After the induction, we may wonder how they could have puzzled us.

The close relation between falsity and partial truth impressed later philosophers such as Benedict Spinoza and G. W. F. Hegel. They were thinking of the truth of entire worldviews, however, so it is likely that their insights may have been independently gained. The Baconian use of the idea is modest and local. Its use does not replace our notions of knowledge, reality, logic, or truth. We assume all that from classical philosophy. Francis Bacon's device is used merely to extract a schematism of the whole of a tiny bit of nature. Of that little bit, we want to understand all its facets in one schematism. We can now appreciate why Baconian induction begins modestly at the base of the pyramid. When attempting to understand the whole nature of something, we must restrict the scope of our study to something manageable. If we want to know and not just guess, it is wise not to bite off more than we can chew. To ask for the whole truth of everything all at once is too much. Even modest goals about a local reality may be unreachable at times.

The schematism or model Francis Bacon asks us to seek in solving a puzzle is impossible to guess in advance. The latent or hidden configuration and its latent or hidden process are unimaginable when we begin. That is why science is needed. At the point of discovery, the model may astonish us.

> The *latent process* of which we speak is a very different thing from anything that will occur readily to men's minds.... For we do not mean actual measures, signs or stages of a process that is visible in bodies, but a wholly continuous process that for the most part escapes the senses. (Bacon, II.6)

We are back with Meno's riddle. We need a mark by which to recognize the true Idea. Without a criterion to pick it out, we would have no reason to seek knowledge. But now we know that mark: the key to the puzzle is the only *consistent* account of the *thing's whole nature (all the distorted appearances we have patiently assembled)*. We may still be unconvinced until it has given us power over nature. If we could learn to manipulate things, we would know we have got hold of them. But Francis Bacon's natural history is intriguing in this way: the hidden key we discover also shows us how to manipulate it. The two achievements, he says, are very nearly the same. We can see how by constructing a simple schematic table.

Let Cs represent the conditions of observing the errors Es.:

Trials		Proximate Trials	
C1a	E1a	C1b	E1b -> -E1a
C2a	E2a	C2b	E2b -> -E2a
C3a	E3a	C3b	E3b -> -E3a

C1a is the condition under which we notice the apparent nature E1a. In a proximate condition C1b, the nature of the thing being studied exhibits itself as E1b, which denies the nature E1a. So there are possibly errors in the evidence on either side of the table. The following line is another similar pair of entries, and so on, giving us the tables of presence and absence of the apparent natures.

When we solve a complex Baconian puzzle, we will have discovered a schematism or a model from which we can derive each error (one of the E's) from its observed condition (the C's): in condition C1a, now the causal condition, the thing appears to appear erronously as E1a, now called the effect. Change the condition slightly to C1b, and you get the effect, E1b. You get more powers as you go on down the list. All this is found in the model, which constitutes the scientific advance.

A detailed interpretation of the nature of a thing by correcting its distortions has given us power over it. If we want to create an effect, we know how when we have recorded it as a discrepant fact. We know which condition will produce it. All we need is a consistent schematism in which all the phenomena are normalized or adjusted to be mutually compatible.

Francis Bacon took the precaution of calling the new middle axioms a "first harvest." (Bacon, II.20). As we climb the ladder up the pyramid, we will no doubt find new discrepancies between the established middle axioms. Further experiments may refute our models. Idols missed will be caught later. We will be redeemed, for a consistent account of the whole of reality will take many trials and many levels of error detection.

Even when we err, his method is successful. We will have discovered a Baconian schematism. Each middle axiom or principle found by his method, even if not the true Idea of the thing in Plato's sense, gives us a limited new power over nature. It fulfills his Christian aim of charity. The new power gained by a discovered Baconian model is also a "pledge of its truth."[2] It is not, to be sure, a criterion of its infallible truth; we are too skeptical to allow that. It is a pragmatic sign that we have "taken hold of the thing" (by a *katalepsis* or a mental grasp). We can assure ourselves that we are not merely playing verbal games with one another as we would have been in a scholastic disputation.

Notes

1. There is a similiairty between Baconian induction, if we neglect its non-monotonic feature, and the deductive method of eliminative induction, which could explain the popularity of the deductive intepretation. Prominent among the deductive interpreters of Baconian induction are Charles D. Broad (1926), Lisa Jardine (1974), L. Jonathan Cohen (1980) and Stephen Gaukroger (2001).
2. Francis Bacon's method may be forgotten by many, but his idea of a pragmatic pledge of truth has found an echo in philosophy of science. Charles S. Peirce was the first philosopher of science to note that modern science provides pragmatic benefits, leading him to suggest that knowledge and truth are closely tied to pragmatism. Ian Hacking (1983) suggested that, unlike theorizing, experimenting gives us confidence in a real entity when we can manipulate it. James Woodward (2003) has proposed that the manipulation of postulated entities is central to understanding causal explanation in science. However, when we look for Francis Bacon's work as part of the "maker's knowledge tradition" (Pérez-Ramos 1988) it is a stretch. Francis Bacon contrasts epexriments that enlighten from those

that are fruitful. His call for science is based oin his observation that fruitful experiments (for instance in natural magic, or in alchemy) fail to enlighten, whereas luciferous exprriemtns are by their nature also fructiferous.

References

Bacon, Francis. (1603) 1857. Valerius Terminus of the Interpretation of Nature: With the Annotations of Hermes Stella. In Vol. 3 *of The Works of Francis Bacon*, edited by James Spedding, Robert Leslie Ellis, and Douglas Denon Heath. London: Longman & Co.

———. (1620) 1858. Novum Organum. In Vol. 4 *of The Works of Francis Bacon*, 14 vols, edited by J. Spedding, R.L. Ellis, and D.D. Heath. London: Longman & Co.

———. (1609) 1884. The Wisdom of the Ancients. In *Bacon's Essays and the Wisdom of the Ancients*, ed. Basil Montagu. Boston: Little, Brown, and Company,

———. (1601) 1951. The Masculine Birth of Time. In *'Temporis Partus Masculus: An Untranslated Writing of Francis Bacon'*, translated by Benjamin Farrington. *Centaurus* 1 (3): 193–205.

———. (1597) 1985. The Essays. In *The Essayes or Counsels, Civill and Morall*, edited by Michael Kiernan. Vol. XV of *The Oxford Francis Bacon*. Oxford: Clarendon Press.

Broad, Charles D. 1926. *The Philosophy of Francis Bacon*. Cambridge: Cambridge University Press.

Cohen, L. Jonathan. 1980. Some Historical Remarks on the Baconian Conception of Probability. *Journal of the History of Ideas* 41 (2): 219–231.

Crawford, Peter. 2014. *The War of the Three Gods. Romans, Persians and the Rise of Islam*. New York, NY: Sky Horse Publishing.

Einstein, Albert, and Leopold Infeld. 1938. *The Evolution of Physics*. London: Cambridge University Press.

Gaukroger, Stephen. 2001. *Francis Bacon and the Transformation of Early-Modern Philosophy*. Cambridge: Cambridge University Press.

Hacking, Ian. 1983. *Representing and Intervening*. Cambridge: Cambridge University Press.

Heather, Peter. 2018. *Rome Resurgent: War and Empire n the Age of Justinian*. New York, NY: Oxford University Press.

Jalobeanu, Dana. 2015. *The Art of Experimental Natural History: Francis Bacon in Context*. Bucharest: Zeta Books.

Jardine, Lisa. 1974. *Francis Bacon: Discovery and the Art of Discourse*. Cambridge: Cambridge University Press.

Joseph, H.W.B. 1916. *An Introduction to Logic*. 2nd ed. Oxford: Clarendon Press.

Kuhn, Thomas. 1962. *The Structure of Scientific Revolutions*. Chicago: University of Chicago Press.

Lai, Tyrone. 1989. How We Make Discoveries. *Synthese* 79 (3): 361–92.

Pérez-Ramos, Antonio. 1988. *Francis Bacon's Idea of Science and the Maker's Knowledge Tradition*. Oxford: Clarendon Press.

Pesic, Peter. 2000. The Clue to the Labyrinth: Francis Bacon and the Decryption of Nature. *Cryptologia*.

Popkin, Richard. 1960. *The History of Skepticism from Erasmus to Descartes*. New York: Humanities Press.

Popper, Karl. (1963) 2002b. *Conjectures and Refutations: The Growth of Scientific Knowledge*. London: Routledge.

Quine, Willard Van Orman. 1951. Two Dogmas of Empiricism. *Philosophical Review* 60: 20–43.

Woodward, James. 2003. *Making Things Happen*. Oxford: Oxford University Press.

9

A Method for Mathematical Physics

I

It is not uncommon among scientists reflecting on the history of science to give credit to Galileo Galilei for bringing the methodology of experimentation to prominence in modern science. He was experimentally inclined. He was brilliant at manipulating materials used in his experiments. His telescopes, for instance, were built for the specific purpose of discerning distant objects in the sky. He had pioneered the use. Great experimental skill was required to construct telescopes suitable for astronomy. No one else at the time could make them well enough to see the moonscape or sunspots as he could, nor other objects in the sky. At least none were made for many years that could serve as well as his better telescopes. Galileo Galilei also practiced his experimental craft with great skill in different ways: with inclined planes, gunnery design, falling bodies, measuring instruments, mathematical compasses, pendulums, and more. His talent made it natural for later commentators to see him as the one person more than any other who introduced the experimental method to modern natural philosophy.

J. Hattiangadi, *Francis Bacon's Skeptical Recipes for New Knowledge*,
https://doi.org/10.1007/978-3-031-52585-8_9

His contribution needs to be reviewed. When he wrote about it in his book *Dialogue Concerning the Two Chief World Systems*, the methodology he recommended did not give as much importance to experimentation as a method to obtain knowledge as his brilliant experimental craft had exhibited (Galilei 1953/1632). If we examine that work closely, we find that Galileo Galilei, as a methodologist, did not trust experiments to generate knowledge or science. An experiment may be used to suggest what knowledge lies beyond deceptive appearance, after which we may want to pursue it. Experimentation, however, does not or cannot provide that knowledge. Francis Bacon, who also mistrusted observations and experiments, also proposed his method of the "interpretation of nature" to cope with their untrustworthy character. The interpretation of nature was proposed to provide little bits of genuine knowledge to build them up, bit by little, into the true pyramid of reality that he thought would become known to us. As they were applied in their later historical evolution, the two methodologies first intertwined and later interfered with one another. The thought of philosophers during the latter part of the seventeenth century was a combination of the two. But each combined them differently from the other, which led to some interference, where one impeded the other. They interfered with each other even more in later centuries. This interference is an integral part of why Baconian induction was forgotten as a theory of making philosophical breakthroughs. This amnesia was evident even while the experimental philosophy continued to be practiced as a craft simultaneously. This strange historical development requires us to study Baconian induction and Galileo Galilei's thoughts on method side by side as independent philosophical contributions concerning human knowledge or epistemology, which is what they were. Only then can we appreciate how later the interference between these outstanding contributions to modern philosophy has influenced epistemology.

II

Francis Bacon's contribution to the theory of knowledge and the related idea of induction addressed some Aristotelian concerns in *Posterior Analytics*. He contributed to a revision of epistemology and a revised form of Aristotelian *epagoge*, or induction, by using techniques derived from various forms of skepticism that are to be found in Plato's dialogues, as they were interpreted in Plato's Academy, and to some extent also by Pyrrhonians. Francis Bacon admitted to the truth of a very dark, pessimistic form of skepticism from which no escape seems possible. To this dark estimate, he made a significant optimistic contribution, which he called his method of "true" induction (Bacon, I.14 and 40).

He frequently contrasts his method with Aristotle's, calling his way a new kind of induction. His contribution was a breakthrough in the theory of knowledge and the approach of the method available to us of gaining it. "True" induction, or the "experimental philosophy" as it was called when employed with some success by the Royal Society since 1660, is the fruit of an epistemological inquiry. For example, Robert Boyle and Isaac Newton were two self-proclaimed philosophers who practiced experimental philosophy during that period, and they were spared the need to address fundamental epistemological questions. Since then, natural philosophers as experimentalists did not need to address fundamental epistemological issues because they presumed them settled by Francis Bacon.

Aristotle proposed that knowledge differed from the merely true opinion as follows: when we know (or understand) something, we know why it is so, and why it could not be otherwise (Aristotle, *Posterior Analytics*, 71b).[1] A demonstrative syllogism is one path to knowledge. When we derive what is to be shown from a valid demonstrative syllogism, we know why it is so (i.e., the cause, or "because" of it, is given by the middle term in the premises). We can understand why it could not be otherwise from the logical necessity of the conclusion, given the premises in a valid argument. A mere opinion lacks this feature of truth that follows necessarily from the premises.

The logical necessity of the conclusion of a valid argument, however, is known provided the premises are demonstrated. Aristotle had shown that a false premise may also be used to validly derive a true conclusion. Aristotle argued that a valid syllogism by itself is not enough. His observation is based on the existence of two distinct kinds of propositions. He argued as follows. Among statements that we entertain, some opinions happen to be true, though not known to be true, and there are also statements known to be true. Therefore (surprisingly) we must allow that there is some form of proof other than that obtained by a demonstrative syllogism.

Knowledge implies the necessity of the thing known. A dermonstrative syllogism requires demonstrated premises to be make the conclusion necessary. There only two possibilities for this to hold. Either we demonstrate the premises using new and different premises, or we do not. If we use new premises, then we need new demonstrations for their premises, in turn. Demanding new premises each time we find ourselves in that famous vicious "infinite regress." Since we cannot provide an infinite number of syllogisms to buttress any knowledge claim, we will have to renounce all claims to knowledge. Then there is only opinion, no knowledge, contrary to our supposition.

If we allow a previously demonstrated judgment as a premise, we have a circularity ("reciprocity"). Any opinion constitutes knowledge if we allow circularity since any opinion can be demonstrated assuming its truth. But we began by asserting that true opinions and knowledge exist and that they are distinct. But after a careful analysis it seems that either every opinion is known to be true, or none are. There must therefore be non-demonstrative knowledge of first principles for there to be a distinction between opinion and knowledge. (Aristotle, *Posterior Analytics*, 72b). "*Epagoge*" or induction, according to Aristotle is how we get to know first principles non-demonstratively. Up to this point, Francis Bacon does not object.

To generate the necessity of first principles, however, Aristotle suggested that a direct grasp of the mind provides it. Aristotle had recourse to a discovery of Plato's to show how this can give us necessary truth. Plato discovered that the Forms are related to one another by a special kind of necessity. Aristotle made use of this insight.

Aristotle departed from some details of Plato's theory of Forms, holding that Forms were embedded in things that change. Plato frequently described the Forms of Ideas as existing apart from those they resemble. An essential attribute of a thing determines its natural or characteristic motion. Aristotle's distinctive philosophy was to lodge the Form in the changing thing. He regarded individuals as possessing primary being, or substance. They exist on their own. Their attributes have secondary existence. Their being is dependent. For instance, Aristotle was a real individual. He was a man. (Being a man is attributed to him.) But the form of Man does not exist indpendently. Being a man is only an attribute of Aristotle and other individuals like him.

Primary beings or substances can support contrary attributes. This apple, which was once green, has turned red. It supports these contrary attributes. "Green" and "red" cannot in turn support contrary attributes. Green cannot become a shape, or a number, or life form. Green is necessarily a color. The necessity of Forms is different—it is not a demonstrative necessity. Individuals support contraries in time. Nothing is green and red at once, so contraries must be attributed in time. The color green, however, is timelessly a color and no more or less now than at any other time.

Aristotle thought that of the many predicates that may be truly attributable to a thing, only one of them is true of it such that we say "that is what it is." When I see a person, Socrates, I see him as a person, a human being. I see him for what he is. Walking unshod in Athens may be true of Socrates at times. But being unshod is not the attribute by virtue of which Socrates is what he is. On another day he may appear well-shod, as he walks towards a celebratory feast. When an attribute is not its essence, or the Platonic Form embedded in the thing, its true nature, it will fail the test of steadfastly belonging to the thing. Anything more green and less at different times is not essentially green. An attribution could be otherwise if it accidental. Socrates is a rational animal because he is a man and men are necessarily, but non demonstratively, rational animals.

There arises the question about how we obtain this knowledge of essence from our perceptions. How do we get to know that a human being is an animal (*genus*) that is able to reason (*differential attribute*) which presumably we perceive? On this important point, Aristotle's writing is unclear, or incomplete. We can gloss over it by saying that we obtain such knowledge from perceptions by a process of abstraction. We look at all the attributes of the species of which an individual is a member. From this list of attributes, we winnow out all the unsuitable attributes until we are left with the last predicate standing. The last predicate will be the kind or species of the thing defined by the method of "genus per differentia." Its attribution is necessary, because it could not be otherwise. Such an understanding of what a thing is, as such, is not that far from our common understanding of things. Aristotelian essential attributes often correspond to concepts of our everyday language. This may give us a sense of familiarity and confidence even though it is not easy to say what Aristotle's method is when we get down to its details.

The nature of a thing (or its *eidos*, its Idea, its species) is only one of its many attributes, but this attribute is unique as its nature among its attributes. When we produce demonstrative knowledge, we must rely on induction to generate first principles that *uniquely* define the nature of the thing to be understood. To know the first principle of a thing is to apprehend mentally (with the *nous*) what it is in itself. There cannot be two natures for something (not two Ideas, not two species.) To have knowledge of the nature of a thing we have to find the one attribute that constitutes its real nature. Francis Bacon uses this framework in his theory of inductive knowledge.

Francis Bacon objected to Aristotle's account of the method. An intuitive grasp of natures by simple perception supplemented by mental abstraction struck him as too naïve and too prone to error. To replace it, Francis Bacon proposes adopting a dialectical analysis like a cross-examination of the nature of things (Bacon 1858/1620, 263). As we have seen, this extension of Socratic dialectics was his new method. Implicit in it is a very different understanding of knowledge.

Francis Bacon was very impressed with some alchemical discoveries. Whereas Aristotle had relied on common observations to yield knowledge of the first principles of perceptible things, alchemists countered with a different conception of reality by relying mainly on little-known, esoteric experiments. They claimed their experiments revealed a hidden and extraordinary reality, quite different from Aristotle's humdrum world. Francis Bacon wanted to use the discovery of strange properties not to reveal anything directly but to challenge the humdrum Aristotelian properties of familiar things.

Francis Bacon's central insight into the theory of knowledge is that we should try to become more aware of our errors in our best claims to know. We need not be worried by the most extreme form of skepticism he could imagine, whereby the mind is by its constitution full of error. We should nevertheless note all the errors we can find. We should not bother looking for a class of pure thoughts, or error-free propositions, even in reports about our perceptions on which we can entirely rely. He rejects all "safe harbor" strategies, such as Locke's simple ideas of sensation (which were proposed later) because the dark truth is that the mind is like a crooked mirror that distorts everything it apprehends. Moreover, he says the mind is unlike writing on a wax tablet, on which we must wipe out the old to write on it anew. It is different with the mind: we must write on it first to wipe out the old, as he says at the end of his fragment, "The Masculine Birth of Time" (Bacon 1951/1601).

The metaphor of mental lenses will simplify and summarise Francis Bacon's method, which will help us to compare it to Galileo Galilei's way. Whatever people decide is truly real may be captured as "what their mind sees," whether the seeing is sensory, perceptual, conceptual, or philosophical. Suppose that seeing reality is inevitably through a distorting lens.

Had we just one distorting lens to contend with, the mind would be trapped in its distortion. We must take reality as seen. We have no independent means of knowing that it is distorted. Fortunately for us, we have many distorting lenses of the mind—many different kinds of lenses and sources of distortions, as it were. We can then examine one image of reality against another. Both cannot be real since nature is unique among forms. A discrepancy arises in the deceived mind, which is our clue.

A mind uses varied lenses, each distorting reality in this way or that. There are too many to enumerate, but we may classify them. Francis Bacon calls the distortions "idols" reminiscent of Plato's *Eidola* or false Ideas. (Bacon, I.41–42). As a tribe of human beings, for one thing, we are prone to a host of different errors. Plato noted many in the *Theaetetus*. Later skeptics added more. Idols can be classified if we wish, but the important thing about them is their immense variety.

Francis Bacon suggests that this wealth of distortions at our disposal is of great value to us. We have a resource, he says, that we have not adequately utilized, and he will show us how to put them to use. Francis Bacon suggests that this wealth of distortions at our disposal is valuable. We can use any one distortion to exhibit the error in another. A lens that shows him taller and thinner is countered by one showing him stouter and shorter. Which is he? Our bounty is a boundless number of Idols. We know that we have erred by the discrepancies in what we see. Any Idol that distorts the mind is threatened—by other Idols! When we know that we have erred there is hope.

Francis Bacon suggests the method of "true" induction, which is ingenious. We take one small item to study—e.g., one species or a quality predicated of things in some closely related species. It is important that the object of study be carefully circumscribed in its scope, to keep the project simple, because it can get out of hand if we do not. The recommendation to perform a "true" induction now is to write up what he calls an "experimental natural history" or a historical record of the various distortions of the nature of a thing that appears before the mind. In his kind of natural history, we note the errors or distortions we can detect by playing up one distorting lens against another. An experimental account is unlike a typical scholastic natural history, where we find typical specimens, i.e., those that reveal and do not distort the underlying form.

Francis Bacon's makes two provisos for an experimental natural history. First, we must note down *all* the distortions (errors) we can find, and second, in each case note the conditions under which we observe the errors. The conditions must be noted clearly. They must be clear enough that the described experiment can be replicated. Any outlandish fact is welcome. Such facts challenge our humdrum expectations. But one may become a part of an experimental natural history only when the conditions of observation have been adequately described.

When writing up an experimental and natural history, it is imperative not to leave something out because it could be explained away. Errors can always be explained away. The mind naturally wants to soar to the highest levels. If we adopt an older metaphysical hypothesis, we can always explain away any discrepancy. What the mind needs are "not wings but leaden weights." For we know already that if we take each part of each discrepancy, one by one, it can be dissected and analyzed away by any philosophical system. By deflecting potential criticism to some minor details of the system, the main system can be propped up as long as we wish, thereby perpetuating the error.

Getting beyond the distorted appearances to the undistorted reality, i.e., behind the distorting lenses, is not easy and always uncertain. But that is when an induction is "true," or faithful to undistorted reality. Francis Bacon gives us this method as a "help" to the mind (Bacon, I.2

and 9): First, we build up a natural history that is as extensive as possible; then we devise a model or schematism for just this thing we are studying. We seek a reconstruction of the real thing that can explain why, under the noted conditions in our historical record, the correspondingly noted errors or discrepancies arise as they do. When we seek such a schematism or model, we must make sure that we do not rise to the highest level of metaphysical theory. Instead, we try to provide a schematism or model for just this little thing we have chosen to study. What we provide for it is not a global understanding of reality, but a bit of local error analysis to give us knowledge in this temporarily isolated subject. Grand metaphysical notions are abandoned to allow us to focus on the error analysis at hand. For adornment, says Francis Bacon, the metaphysical supposition is better. But for discovering reality, true induction recommends itself.

We seek a model (a combination of the hidden or latent *structure* and its *process*, to use Francis Bacon's words) that can explain why the errors in our little natural history happen as they do, under each of the noted conditions (Bacon, II.1). When we do, it is like solving a riddle, like finding one's way through a maze, like the deciphered code in a secret message, or like finding an object that has been hidden. In each case, the solution is unique if the record of natural history is extensive enough. It is like "puzzle-solving" (Kuhn 1962). "Puzzle" is a useful word. A recreational puzzle proposed by the poser of the puzzle is solved when we have worked out its unique solution. Puzzles for amusement are designed to have unique solutions. If published in a periodical, the correct answer is given in a later issue, demonstrating this unique solution. Natural histories are our way of creating a puzzle for ourselves that seems at first almost impossible to solve. Only one solution will be possible when an experimental natural history is extensive enough. The unique solution, obtained by an error analysis, must be true because it is uniquely coherent.

Let us consider once again Aristotle's requirement that for *our knowledge* to be different from *our true opinions*, we require the former to be not merely true but necessarily so. Aristotle appealed to metaphysical necessity, for it is ultimately Platonic metaphysics that yields us the necessities of Forms he relies upon. It made Aristotle's epistemology vulnerable: his theory of knowledge cannot say how we come by the ontic knowledge without circularity. We may ask, which is first: his metaphysics or his epistemology? His opinion on method was easily countered by skeptics as "dogmatic," i.e., for substituting opinion (about Forms) for knowledge.

Francis Bacon's epistemology is superior: the epistemic necessity in a Baconian principle is in turn epistemic. A natural puzzle is born in a nest of impossibilities of apparent natures. A solution, when found, is unique: any alternative construal of the facts is impossible. Necessity derives partly from the impossibilities summarized in the unsolved puzzle and partly from the uniqueness of the solution. It is necessary because nothing else is possible in the case at hand once we get to appreciate the puzzle that generated it.

Francis Bacon makes another case for the "truth" or faithfulness of discovery by error analysis: *power over nature*. The schematism or model solution tells us, for each discrepancy, how a different condition gives rise to its distortion. If we call the distortions "discrepant effects," and the conditions under we which we note their respective occurrence their "causal conditions," then we know from the solution exactly how to produce the effects—namely by producing the same causal conditions that we have noted in our natural histories. This is how induction begets "power over nature." It is often noted, with a nod to Baconian thought, as a feature of our manipulation of nature or of our intervention into it, which is conceived as an adjunct to the hypothetico-deductive method (Hacking 1983, Woodward 2003). But gaining a new power over nature is more central to Baconian induction. Any solved puzzle provides a schematic structure showing us how to manipulate nature.

Galileo Galilei had a different method in mind.

III

Galileo Galilei was a mathematician by profession, and an exponent of the techniques of Archimedes, whose mechanics he taught. He was also a keen philosopher, generally speaking, and a metaphysician. He had attended undergraduate lectures on philosophy in the Collegio Romano. All recent comments on his work by historians, social theorists, and philosophers of science acknowledge his philosophical interests (Koyré 1978.) But there is a wide variety of opinion concerning Galileo Galilei's contribution to modern science, about his method, and how his work should be interpreted. Some have said that since he followed the Archimedean method, he must have been a Platonist, like Archimedes. Others read his early writings and concluded that he followed Aristotle's method. Yet others claim that he was following well-established medieval ways, different from both. Yet others that while he was an Aristotelian about the understanding of experience, he was much more influenced by the skills of the manual tradespeople, in whose work he took a keen interest. These historical interpretations of careful scholars seem to be partially correct in the following sense. If a proponent of one of the views does not say that one influence on it is all there is to Galileo Galilei's thought, then we can readily acquiesce. Unfortunately, that is often not the case, so we cannot consent.

Pierre Duhem, who was Galileo Galilei's most influential modern detractor, recognized a metaphysician in him. To preserve the autonomy of physics as science from such metaphysics or philosophy, he asked us to regard his contributions to physics as uncontroversial, small, and incremental. They were useful only so long as we ignored his metaphysical opinions. Pierre Duhem thought that when we take away the metaphysical interpretation of physical theory from it, what is left behind is a combination of mathematical technique and a bank of experimental data, both of which always grow very slowly, in minute increments, with the assent of all concerned (Duhem 1954/1906, 7–18).

Alexandre Koyré called out Pierre Duhem in defense of the scientific revolution of the seventeenth century, "...what the founders of modern science, among them Galileo, had to do, was not to criticize and to combat certain faulty theories, and to correct or to replace them by better ones. They had to do something quite different. They had to destroy one world and to replace it by another." (Koyré 1943).

"... [An] unbroken tradition leads from the works of the Parisian Nominalists to those of Benedetti, Bruno, Galileo and Descartes.... Still the conclusion drawn therefrom by Duhem is a delusion: a well-prepared revolution is nevertheless a revolution, and in spite of the fact that Galileo himself in his youth (as well as at times Descartes) shared the views and taught the theories of the medieval critics of Aristotle, modern science, the science born from his efforts and discoveries, does not follow the inspiration of the 'Parisian forerunners of Galileo'; it places itself at once on a quite different level—on a level which I should like to call the Archimedean one. The true forerunner of modern physics is neither Buridan, nor Nicole Oresme, nor even John Philoponos, but Archimedes." (Koyré 1943) Archimedes was, of course, a great admirer of Plato. So was Galileo Galilei. We find Alexandre Koyré concluding his paper by describing Galileo Galilei as a Platonist. "The new science is for him an experimental proof of Platonism." (Koyré 1943, 428).

Galileo Galilei admired Plato, as any good Archimedean mathematician would have done. Furthermore, he had read Plato's dialogues closely. He had even applied Plato's thought in many cases he was examining wherever it seemed appropriate. But Platonism is not all there is to Galileo Galilei's philosophy. He was, on the contrary, very much his own philosopher.

When Galileo Galilei was offered a job as a mathematician in Florence in 1610, he insisted that his new title be philosopher and mathematician to the court. To find out what he means by "philosopher and mathematician" we can turn to Plato, who had something relevant to say about knowledge, mathematicians and philosophers in the *Republic*.

In a manner reminiscent of Parmenides, Socrates in Plato's *Republic* maintains that knowledge pertains to "that which is," or Being, whereas what he calls nescience pertains to "that which is not." Opinion pertains to what is and is not, or to Becoming (Plato, *Republic*, 476d-480a). The phrase, "that which is and is not" is pivotal here. Everything that changes, in time, will be first one thing and then another thing that is not what it was earlier. Realities (e.g., the Forms,) of course, do not share this feature. They just are. Hence, we can know the Forms, because they are intelligible, whereas that-which-is-and-is-not is very hard for us to grasp clearly, and it is not intelligible. A thing changing is uncertain (in Heisenberg's sense of being smeared.) Whatever we say about that-which-is-and-is-not cannot be known to us but may be perceived, and will always remain a matter of opinion, thought Plato. This difference is clarified later in the *Republic* in his famous Analogy of the Divided Line, which will help us understand what Galileo Galilei meant when he said he wanted to be both a philosopher and a mathematician (Plato, *Republic*, 510d-511e).

A line is divided, using a principle of proportion, into four unequal parts. The first section in each half of the line bears the relation to the next in that half that the sum of the first two parts bears to the sum of the other two. The conditions are fulfilled of the parts of the line are gradually descending in order of their length. Plato is not clear if the lines in the analogy are decreasing or increasing as we go along, so we must make a choice here. The significance of proportionality is not clear, but we can ignore what its meaning might have been for Plato.

The largest part of the line, the first, represents images and shadows (we could add sense impressions to them, perhaps). Images occupy our minds the most. The second longest segment represents what we know of the changeable objects themselves. Things are clearer and more permanent than their fleeting images. The things of the second part of the line thus reach into and make sense of the objects of the first part. However, we can only find objects of perception in these two sectors of the line. They cannot be real, but only apparent knowledge. We cannot have exact knowledge of anything apparent and time-bound since they change in time. Exact knowledge of things requires that they must be exact enough in themselves to be grasped exactly by our minds. Plato could not understand how the objects of perception, which are smeared, could ever be known in that manner.

When we cross over to the next two (smaller) parts of the line, they represent knowledge of changeless abstractions. On the third largest part of the divided line are those intelligible things treated by geometers and reckoners, i.e., by mathematicians. Things in this section of the line compel us to use perceptible things (from the second part of the divided line, of course) to understand the intelligible forms of mathematics studied therein. This means that we use diagrams to make sense of geometrical proofs.

Moreover, geometry does this: "by means of assumptions… it proceeds not up to a first principle, but down to a conclusion…" (Plato, *Republic*, 510b). In short, the method of mathematics is to assume hypotheses to be clearly understood and true, and then deduce its consequences, shedding light on perceptible things that resemble the intelligible things being studied.

Mathematicians cannot investigate their first principles' meaning and truth. They merely formulate mathematical hypotheses and explore their consequences to fit the facts perceived among the changeable objects. The method of hypothesis was called "the method of saving the appearances" when applied to phenomena, as in astronomy. Mathematical objects are exact and immutable. We can acquire knowledge of them for that reason. However, our mathematical knowledge is unfortunately only hypothetical. When it is applied to changeable objects to shed light on them, the knowledge becomes approximate and unclear.

Plato's theme is well illustrated in his dialogue, the *Timaeus*. Timaeus warns Socrates from the beginning that knowledge (*episteme*, or science) of celestial motion is impossible because it is motion. Knowledge can only pertain to unchanging entities, which are those kinds of objects of interest to mathematicians (Plato, *Timaeus*, 27d-29d). Timaeus has an opinion about celestial motions, which he goes on to develop in the dialogue, which may be called a "likely story." It contributed to Eudoxan astronomy using the method of concentric circles to capture some errant appearances of planets. The best we can do in astronomy and related cosmology and theology, with mathematics, is *to save the phenomena*. Mathematicians cannot know what stars are or what paths they really follow by using mathematics alone. Mathematicians do not inquire into the truth of their own assumed first principles. Many mathematical models are possible. In studying motions, the best that a mathematician can do is to save the phenomena.

In his passage on the Analogy of the Divided Line, there is no special mention of the Ideas or Forms. We may consult the *Phaedo,* in which the theory of Forms is introduced as a hypothesis. We may conclude that Forms, too, are represented in the third largest section of the divided line as hypothetical knowledge. What we know about things is itself hypothetical when we invoke some Forms to understand them, just as much as are all mathematical accounts of celestial objects. Suppose we understood the form of justice reasonably well after a Socratic examination. The light it sheds on the laws of the city of Athens, Sparta, or Syracuse would be hypothetical and not quite unconditionally *episteme*.

The fourth section of the divided line, the smallest, represents the greatest clarity and highest level of abstraction. It gives us "that which reason itself lays hold of by the power of dialectic, treating its assumptions not as absolute beginnings but literally as hypotheses, underpinnings, footings, and springboards so to speak, to enable it to rise to that which requires no assumption and is the starting point of all…." (Plato, *Republic*, 511b-c). Following the Socratic dialectical method, philosophers can find true and unconditional principles at the highest level of all. They rise above merely hypothetical models that save the phenomena. In short, mathematics, and even the theory of Forms, can give us only hypothetical or conditional knowledge of perceptual reality, when we need unconditionally known principles. Only the philosopher can do this.

Galileo Galilei wanted to be regarded as a philosopher because he wanted to rise to that level, knowing "that which requires no assumptions and is the starting point of all…." He wanted to offer a new science of kinematics, not just some hypothetical opinions concerning motion. However, Plato was quite clear about this. The realm in which motions are perceived is not capable of supporting knowledge. There can be no science of local motion, or kinematics, for a Platonist. Galileo Galilei knew his Plato well. He was a philosopher in his own right. He gave us, not Plato rehashed, but the rudiments of modern philosophy. This philosophy made modern mathematical physics possible. Although Alexandre Koyré sums it up as Platonism, we can find isolated comments by him on Galileo Galilei in which Alexandre Koyré seems to recognize that he was also revising Platonic philosophy.

Galileo Galilei's philosophical accomplishments began with this: he began to treat the line or the curve of the geometer as the trajectory of a body. His geometry describes real local motions. It describes change, and not changeless entities. This is a new interpretation of geometry that had no precedent in ancient Greek thought. This what he means by the very expression "science of kinematics." He also expresses this in 1623 somewhat obliquely, "Philosophy is written in this grand book, which stands continually open before our eyes (I mean the 'Universe'), but cannot be understood without first learning to comprehend the language and know the characters as it is written. It is written in mathematical language, and its characters are triangles, circles, and other geometric figures, without which it is impossible to humanly understand a word; without these, one is wandering in a dark labyrinth." (Galilei 1957/1623, 237–8).

With his new interpretation of geometry, he laid the groundwork for the metaphysics of motion, to borrow a phrase. If you doubt that this new interpretation of geometry makes modern physics possible, please try to imagine what modern physics would be like if it could not treat local motions as geometrical lines. If motions were not treated as objects in their own right, we would not have "laws of motion," at all. All laws would pertain primarily to ideal forms or essences, and the motions each of them entails for its kind. Kinematics requires Galileo Galilei's declaration of independence on behalf of all local motions from the hegemony of ideal forms and essential qualities.

When Galileo Galilei began to treat local motion itself geometrically, he made the geometry descriptive of reality in time, and not just a model to capture temporal phenomena that are inherently unintelligible. He had instantly made local motion more exact and more understandable than anything else in the Universe because the exact study of geometry describes it. What was unintelligible to Plato and to Aristotle, namely motion itself had become the one thing we can know the best and the most exactly—i.e., geometrically. Local motion is better understood even than the essences of things, the embodied Forms in Aristotelian thought. Essences themselves were no longer needed to understand motions. They had become irrelevant to natural philosophy. As Alexandre Koyré might have said, Galileo Galilei had rubbished the qualitative world to transport us to a mathematical universe (Koyré 1957).

It is a matter of plausible speculation that Galileo Galilei gradually came to his new metaphysics of motion between 1602 and 1609 in this manner. He was studying motions on earth using the techniques of Archimedean mechanics, assuming the truth of the Copernican hypothesis. Since the old physics was no longer applicable to moving earth, he began experimenting and creating hypothetical mechanical models for falling bodies, pendula, and natural motions on earth. His investigations were old in philosophy, as Pierre Duhem notes, but new in mathematics.

To his amazement, he discovered that each of his studies showed that models that saved the phenomena did not use motions deriving from Aristotelian forms. The law of fall seems independent of the nature of the body that falls if we make allowance for air friction. It depends only upon distance and time. The tendency of a body to perpetuate its own motion in the absence of opposing forces (or the property of inertia) seems independent of the *nature* of the body. The path of a projectile also seems independent of the *nature* of the object being projected through the air. These hypothetical models seemed to depend only on time and distance as variables. Perhaps, he thought, there is a science of motions as such, depending only upon their metric properties. Once he came to believe that the Copernican hypothesis is very likely true, based on observations through a telescope, though not yet known to be so, a new science of motion seemed to be indicated—a "starry message."

This discovery of mathematical physics, of the geometry of motion itself, was not only a metaphysical discovery, of course. Many insights and techniques follow from it, which are to be found in mathematics, physics, and his distinctive methodology. In his distinctive methodology, he acquires an ability to treat motion itself as an object, allowing us to ask questions about it in isolation from interfering bodies around the motion. Galileo Galilei invented a technique of idealized thought experiments because he allowed himself to think of motion itself in abstraction from the bodies that may constrain it.

Regarding mathematical techniques, his focus on the motion allows him to analyze any motion at any point to show how it may be decomposed into independent components at right angles to one another. This technique was known to mathematicians in antiquity, but it was applied mainly in statics to analyze displacements. The importance of this principle of independence of orthogonal motions for subsequent physical theory, or what may be called the Principle of Superposition, as it is known today, cannot be overemphasized. Related to this is the idea that two motions of a body can be added to yield a compound motion for it, using a parallelogram whose diagonal represents the compound when the sides are drawn proportionate to the motions. Component motions can also be subtracted from a compound motion to calculate the remaining component motion using the same parallelogram method. His method allowed him to ask questions about pendulum motion itself, the law of falling bodies, or the geometry of projectile motion. He could approach physics with a generality that goes beyond any inquiries before he had proposed this new interpretation of geometry.

Galileo Galilei made the many different natures otiose that Aristotle had used to invoke, i.e., the different natural motions of the different kinds of thing. When we can treat motion mathematically through a few qualities, we can banish all forms, species, and qualities into a subjective realm. Geometry, widely applied, provides us with all the Forms we need for physics (these are the primary properties or qualities of things.). In 1623, he writes, "… [W]henever I conceive any material or corporeal substance, I immediately feel the need to think of it as bounded, and as having this or that shape; as being large or small in relation to other things, and in some specific place at any given time; as being in motion or at rest; as touching or not touching some other body; and as being one in number, or few, or many. From these conditions, I cannot separate such a substance by any stretch of my imagination" (Galilei 1957/1623, 274). We note here how close this thought to the reason René Descartes gives for identifying matter with space, though this is not the intent of Galileo Galilei's thought.

We note moreover that he continues as if prompting Cartesianism; "But that it must be white or red, bitter or sweet, noisy or silent, and of sweet or foul odor, my mind does not feel compelled to bring in as a necessary accompaniment. Without the senses as our guides, reason or imagination unaided would probably never arrive at qualities like these. Hence I think that tastes, odors, colors, and so on are no more than mere names so far as the object in which we place them is concerned, and that they reside only in consciousness. Hence if the living creature were removed, all these qualities would be wiped away and annihilated" (Galilei 1957/1623, 274).

Modern metaphysics is unfolding before our eyes. Proposed lists of primary qualities can vary with different theories of motion under examination. It makes little difference to Galilean metaphysics. What matters is this: there is only one kind of motion: it is geometrical in nature. Galileo's metaphysics is called "classical physics" nowadays. Its universal laws of motion create all the great wealth of effects that we can perceive on the otherwise nondescript matter of the universe. For three hundred years there seemed to be no other way to study motion in science except in this way, as classical or modern physics did, from 1623, and until 1927.

Galileo Galilei faced an epistemological challenge. The relations between abstract entities make our knowledge different from opinion.

Geometry gives us knowledge of things, but our sense experience only gives us subjective opinions. When Galileo Galilei banished all forms and qualities of Aristotle's philosophy into a subjective purgatory, he was in epistemological trouble. The source of the necessity of knowledge in the philosophies of Aristotle and Plato was in the Forms. Their theory of knowledge relied on the non demonstrative necessities in the Forms. If Formal truths are merely subjective or qualitative opinion, he has lost the necessity of knowledge. How is it to be distinguished from mere opinion? For his kinematics to rank as a science, or to count as knowledge, he needed a new source for its necessity. Physics, therefore, could not claim to be science, or as knowledge, we must conclude, without a new theory of knowledge.

The ever-resourceful Galileo Galilei resolved this acute epistemological deficiency of his geometry of motion. This statement is found towards the end of the first day of the *Dialogue:* "… [T]he human understanding can be taken in two modes, the *intensive* or the *extensive*. *Extensively*, that is, with regard to the multitudes of intelligibles, which are infinite, the human understanding is as nothing even if it understands a thousand propositions; for a thousand in relation to infinity is zero. But taking man's understanding *intensively*, in so far as this term denotes understanding some proposition perfectly, I say that the human intellect does understand some propositions perfectly, and thus in these has as much absolute certainty as Nature itself has. Of such are the mathematical sciences alone; that is, geometry and arithmetic, in which the Divine intellect knows infinitely more propositions, since it knows it all. But with regard to those few which the human intellect does understand, I believe that its knowledge equals the Divine in objective certainty, for here it succeeds in understanding necessity, beyond which there can be no greater sureness" (Galilei 1953/1632, 103).

Galileo Galilei thought that Plato's Socrates, whom he admired, knew certain wise truths, even though he professed ignorance of wisdom in the *Apology*. This is how his being the wisest (according to the Oracle) and the Socratic profession of ignorance are reconciled. "Socrates recognizes his knowing nothing relative to absolute wisdom, which is infinite…. Socrates did well to recognize his limited knowledge as nothing compared to the infinity which he lacked" (Galilei 1953/1632, 102). His interpretation of the Platonic dialogues is that it is didactic rather than as skeptical. The latter was Francis Bacon's interpretation. There is a long tradition of reading Plato as Didact, and that we are his students and inquirers. This tradition goes back to the neo-Platonists. On this reading, all the dialogues of Plato must be read as teaching us some secret wisdom. The lessons are set out in dialogue form to awake in our mind the need and the ability to see certain wise things that he had discovered, whose key is that we must discover the wise truths for ourselves.

Galileo Galilei's kinematics can be science, too, if it is based on an intuitive certainty. Its necessity arises from its certainty because it is as sure as divine knowledge. Galileo Galilei's bar for knowledge is set high. This new divine epistemic bar gives us a way to generate the necessity for the truth of our first principles. We can extract its necessity from knowing that they cannot possibly be mistaken. Knowledge can arise from the human intuition that rises to a divine level to apprehend some truth with certainty, though this is possible only in geometry and in arithmetic. Galileo Galilei did not provide examples.

In the *Dialogues*, we find that Galileo mentions experiments, but does not rely on them. He treats them as suggestive of underlying laws. As for knowledge, we must rely on proofs based on thought experiments, which we can trust because they show that denying the principle to be proved leads to the incoherence of our thought. We may disagree with his faith in the epistemological value of thought experiments, but we must do so respectfully, given his wonderful use of them in his later writing.

This new epistemology (incidentally, also a new philosophy of mathematics) is just a hint, a beginning. It gives us a sense of direction, but not the details of what and how we can know with such certainty that it can be compared with divine knowledge. Rationalism and empiricism, later erected as the two pillars of modern epistemology, rest upon the Galilean certainty of our mathematical knowledge, which is here being asserted. We need to study the writings of René Descartes to see how Galilean thoughts were developed along these lines.

In 1623, a decade before the *Dialogues Concerning the Two Chief World Systems* was condemned by the Inquisition, Galileo Galilei had already written about his new thoughts in philosophy, but they were contained as asides in a polemical work against the theory of comets proposed by Lothario Sarsi. This polemic in *The Assayer* (*Il Saggiatore*) is of great interest. Galileo Galilei describes comets as optical illusions appearing to be phenomena beyond the moon. He argues that they were sublunary phenomena and were partially illusory. Physicists and astronomers since his time have repudiated this opinion. The evidence of his error about comets was available to astronomers even since the time of Tycho Brahe, who had studied their parallax of motion. Galileo Galilei was familiar with this argument, as he had used it to show that the supernova was a new star located in the heavens beyond the moon and that it was not a sublunary object creating an optical illusion of a new star. Yet he missed Tycho Brahe's similar work with parallax and the comets.

Instead of describing this as an unfortunate error by a great thinker, we should note that he rejects the phenomena of comets as impossible because he keenly understands the motions of celestial objects. He appreciates that they must obey an inertial law, and it is hard to imagine how comets could obey it from what we see. Although this knowledge, which is partial, leads him to make an erroneous judgment, it is an error that only an outstanding physicist can make. Galileo Galilei does not look around to find out what the astronomers tell us in detail, because the physics is so compelling.

Some decades later, Isaac Newton makes a similar mistake with the great comet of 1680. He would not accept John Flamsteed's description of its path. Being a Baconian experimental philosopher, however, he corrects himself soon, taking the appearance of the motion of comets at face value. Instead of eliminating the error based on a prior philosophy, which would have stopped him writing the *Principia*, he notes the error down as a part of the natural history of comets. He finds a resolution to the puzzle of the impossible motion of the Comets. Thereby, he proposed a new system of the world based on a principle of universal gravitational force.

In the course of Galileo Galilei's misdirected argument, however, he describes the nature of the universe and how we perceive it, in a new and exciting way. "It now remains for me to tell Your Excellency, as I promised, some thoughts of mine about the proposition' motion is the cause of heat', and to show in what sense this may be true. But first I must consider what it is that we call heat, as I suspect that people, in general, have a concept of this which is very remote from the truth. For they believe that heat is a real phenomenon or property, or quality, which actually resides in the material by which we feel ourselves warmed. Now I say that whenever I conceive any material or corporeal substance, I immediately feel the need to think of it as bounded, and as having this or that shape; as being large or small in relation to other things, and in some specific place at any given time; as being in motion or at rest; as touching or not touching some other body; and as being one in number, or few, or many. From these conditions, I cannot separate such a substance by any stretch of my imagination. But that it must be white or red, bitter or sweet, noisy or silent, and of sweet or foul odor, my mind does not feel compelled to bring in as necessary accompaniments. Without the senses as our guides, reason or imagination unaided would probably never arrive at qualities like these. Hence I think that tastes, odors, colors, and so on are no more than mere names so far as the object in which we place them is concerned, and that they reside only in the consciousness. Hence if the living creature were removed, all these qualities would be wiped away and annihilated. But since we have imposed upon them special names, distinct from those of the other and real qualities mentioned previously, we wish to believe that they really exist as actually different from those." (Galilei 1957/1623, 274).

"I may be able to make my notion clearer by means of some examples. I move my hand first over a marble statue and then over a living man; to the effect flowing from my hand, this is the same with regard to both objects and my hand; it consists of the primary phenomena of motion and touch, for which we have no further names. But the live body which receives these operations feels different sensations according to the various places touched. When touched upon the soles of the feet, for example, or under the knee or armpit, it feels in addition to the common sensation of touch a sensation on which we have imposed a special name, 'tickling.' This sensation belongs to us and not to the hand. Anyone would make a serious error if he said that the hand, in addition to the properties of moving and touching, possessed another faculty of 'tickling,' as if tickling were a phenomenon that resided in the hand that tickled. A piece of paper or a feather drawn lightly over any part of our bodies performs intrinsically the same operations of moving and touching, but by touching the eye, the nose, or the upper lip it excites in us almost intolerable titillation, even though elsewhere it is scarcely felt. This titillation belongs entirely to us and not to the feather; if the live and sensitive body were removed it would remain no more than a mere word. I believe that no more solid an existence belongs to many qualities which we have come to attribute to physical bodies-tastes, odors, colors, and many more." (Galilei 1957/1623, 275). In his book, *The World, or a Treatise on Light*, René Descartes writes also of feelings as "tickles," which shows that he was firmly in the camp of Galileo Galilei (Descartes, AT XI.6).[2] It is not surprising that *The World* was withdrawn from publication on the news that the *Dialogues Concerning Two Chief World System* stood condemned by the Church. René Descartes' abandoned book was just as Copernican through and through.

Galileo Galilei continues this line of thought. "To excite in us tastes, odors, and sounds I believe that nothing is required in external bodies except shapes, numbers, and slow or rapid movements. I think that if ears, tongues, and noses were removed, shapes and numbers and motions would remain, but not odors or tastes or sounds. The latter, I believe, are nothing more than names when separated from living beings, just as tickling and titillation are nothing but names in the absence of such things as noses and armpits. And as these four senses are related to the four elements, so I believe that vision, the sense eminent above all others in the proportion of the finite to the infinite, the temporal to the instantaneous, the quantitative to the indivisible, the illuminated to the obscure—that vision, I say, is related to light itself. But of this sensation and the things pertaining to it I pretend to understand but little, and since even a long time would not suffice to explain that trifle, or even to hint at an explanation, I pass this over in silence." (Galilei 1957/1623, 276–77).

The sharp separation of the class of physical objects, namely those capable of local motion, from the objects of consciousness (incapable of local motion) is correctly ascribed to René Descartes. It may give us the impression that the cleavage of mind from matter is a gratuitous philosophical adventure of his. René Descartes certainly took this difference to heart and made it a fundamental feature of modern metaphysics by identifying them as two substances or things that exist in some manner independently of one other. Nothing so dramatic as this is proposed in *The Assayer*, where it is merely an observation about different bits of reality and the different names we give them. Galileo Galilei inspired modern metaphysics without inventing it.

IV

Galileo Galilei thought we can have knowledge that is as certain as God's. Still, he did not include the Copernican Hypothesis itself within the scope of what he knew. In the second revised edition of his translation of the *Dialogue*, Stillman Drake as the editor records that the following note is found "added by Galileo in the preliminary leaves of his own copy of the Dialogue.

> Take note, theologians, that in your desire to make matters of faith out of propositions related to the fixity of sun and earth you run the risk of eventually having to condone as heretics those who would declare the earth to stand still and the sun to change position—eventually I say, at such a time as it might be physically or logically be proved that the earth moves and the sun stands still. (Galilei 1953/1632, v)

His trial seems to be very much on his mind here. Also on his mind is the fact that he had been warned off as a heretic by the theologians in the Church in the famous trial by the Inquisition.

Students of rhetoric may see in this remarkable passage a clever inversion of his prosecution by theologians employed in the Vatican. But he was very accomplished at polemics, and this clever passage should not be surprising. Others may see an amusing picture of Galileo Galilei wagging his finger at theologians, who are being treated like naughty schoolchildren, for their assertion of an opinion in astronomy, of which they had little knowledge. The finger wagging, too, is Galileo Galilei's characteristic style. Most of his early professional life was spent as a Archimedean mathematics teacher, not as a philosopher. In mathematics, you can be obviously correct or mistaken in the details of your work, and it is up to a mathematics teacher to tell careless students when the latter make mistakes. Finger wagging is normal. After 1610, Galileo Galilei wanted to be a philosopher and a mathematician. Still, in some of his writing, he retained the finger-wagging style of a mathematics teacher who admonishes young students. He did not stop thinking as a philosopher would, of course. However, the nuanced rhetorical style of the inoffensive philosopher, making observations to the sky, was never his. His writing is full of ire, and it may well be that this ire wounded the feelings of some around him.

Philosophers today who read his note to himself will do so very differently than historians who may enjoy a laugh at his wagging finger. Galileo Galilei is clearly saying two, or perhaps three, philosophical things here: first, he is admitting that he does not know that the Copernican Hypothesis is true; second, that it is also his opinion that he trusts or expects it eventually to be *demonstrated or proved* to be true. A third, perhaps, is an unlikely contrast between knowledge (in physical astronomy) and faith (as it is involved in theological knowledge) which are inverted from their traditional roles. It may be worth our while to pursue what this note implies.

From the fact that it is a personal note to himself that only he and an occasional friend who dropped in could see in his own copy of the *Dialogue*, it is clear that the note cannot have been an insincere argument to persuade the Censors to leave him be, or to the world to convince them that he is right. He was already under house arrest, being looked after by his daughter. Therefore, his avowal that he does not know that the earth moves must be heartfelt and sincere. Nevertheless, he hopes to prove this opinion, which he trusted to be true and would one day be known. How could that happen?

We can know a few hypotheses as well as God, he has claimed, and these are mathematical. The theorems are the foundation of knowledge. We could call it Foundational Reason, provided we distinguish it from the conception of discursive reason in the writings of Plato, Aristotle, or Thomas Aquinas, where its infallibility is not required and not asserted.

There may be some Foundational Intuitions, which, together with logical or physical considerations, could leave us one day with the certainty that the earth moves around the Sun. It is possible that the mathematical truths that Galileo Galilei had postulated as knowable just as well as God's knowledge, could yield us, by using deductive reasoning, more complicated truths about the earth and the Sun. But would the mathematical truths be enough to yield our knowledge of this kind? Or would it fall short and give us merely a likely story, or a probably true opinion, of the earth in motion?

Apparently, it is never enough to have infallible first principles. René Descartes was the first to acknowledge this fact in the last section of his *Discourse on Method* (Descartes 1985.) Even if we have indubitable knowledge of the principles of mathematical physics, it is not enough. They do not provide a model of the solar system. They allow too many models from which to choose. We must adopt a hypothetical model to make sense of phenomena. Galileo Galilei's admission that we do not know that the earth moves is the right conclusion to draw in his conception of knowledge. Unless we adopt a different theory of knowledge, it remains a live possibility that the earth is stationary.

The philosophical thought of René Descartes is familiar to all of us in its outline. It is unnecessary to repeat most of it, but we need to examine some parts related to the new method he introduced, inspired by Galileo Galilei. We will see why, for all their admiration of Galileo Galilei and René Descartes, the Fellows of the Royal Society adopted the experimental method of Francis Bacon, as revised by Robert Boyle.

Notes

1. Aristotle's works are typically cited by their "Bekker number," which is a page number from the standard edition of his works and the letter for the column of text on that page ('a' or 'b').
2. References to and quotes from Descartes' works here will all be to the English translations in *The Philosophical Writings of Descartes* (Descartes 1984, 1985, 1991). Instead of citing years and pages, these references will follow the typical citing of passages by their page number in the Adam-Tannery edition of the *Oeuvre de Descartes* (Descartes 1983), as in "Descartes, AT II.370" for page 370 of the second volume of that edition.

References

Bacon, Francis. (1620) 1858. Novum Organum. In Vol. 4 *of The Works of Francis Bacon*, 14 vols, edited by J. Spedding, R.L. Ellis, and D.D. Heath. London: Longman & Co.

———. (1601) 1951. The Masculine Birth of Time. In *'Temporis Partus Masculus: An Untranslated Writing of Francis Bacon'*, translated by Benjamin Farrington. *Centaurus* 1 (3): 193–205.

Descartes, René. 1983. *Oeuvres de Descartes*, 11 vols. Edited by Charles Adam and Paul Tannery. Paris: Librarie Philosophique J. Vrin.

———. 1984. *The Philosophical Writings of Descartes*, vol. 2. Edited by John Cottingham, Robert Stoothoff, and Dugald Murdoch. Cambridge: Cambridge University Press.

———. 1985. *The Philosophical Writings of Descartes*, vol. 1. Edited by John Cottingham, Robert Stoothoff, and Dugald Murdoch. Cambridge: Cambridge University Press.

———. 1991. *The Philosophical Writings of Descartes*, vol. 3. Edited by John Cottingham, Robert Stoothoff, Dugald Murdoch, and Anthony Kenny. Cambridge: Cambridge University Press.

Duhem, Pierre. (1906) 1954. *The Aim and Structure of Physical Theory.* Originally, *La Théorie Physique, son objet et sa structure.* Translated by Philip P. Wiener. Princeton, NJ: Princeton University Press.

Galilei, Galileo. (1632) 1953. *Dialogue Concerning the Two Chief World Systems.* Translated by Stillman Drake. Berkeley: University of California Press.

———. (1623) 1957. The Assayer. In *Discoveries and Opinions of Galileo*, trans. and ed. Stillman Drake. New York: Doubleday.

Hacking, Ian. 1983. *Representing and Intervening.* Cambridge: Cambridge University Press.

Koyré, Alexandre. 1943. Galileo and Plato. *Journal of the History of Ideas* 4 (4): 400–428.

———. 1957. *From the Closed World to the Infinite Universe.* Baltimore: Johns Hopkins Press.

———. 1978. *Galileo Studies*, tr. John Mepham. *Études Galiléenne*, (1938) Atlantic Highlands, NJ: Humanities Press.

Kuhn, Thomas. 1962. *The Structure of Scientific Revolutions.* Chicago: University of Chicago Press.

Woodward, James. 2003. *Making Things Happen.* Oxford: Oxford University Press.

10

A Foundational Scepticism

I

Not many can claim to have slain the ogre of skepticism. René Descartes slew it not once but twice. Needless to say, each form of doubt he slew has survived its slaying by his hands. One of these monsters was the Pyrrhonian epoché, leading to restraint in any assertion. It was familiar to him in the writings of the famous essayist Michel de Montaigne and his admirer Pierre de Charron. He invented another form of skepticism by himself. It is often called "the method of hyperbolic doubt" or foundational doubt. David Hume called it an antecedent skepticism because what follows it, if successfully deployed, is indubitable knowledge. It is a particular form of skepticism that made René Descartes famous. His foundational doubt involves a single thought experiment that promises a foundation for all scientific knowledge. David Hume called this an antecedent doubt because its purpose is to leave us with a foundation of knowledge that is free of it. Galileo Galilei's suggested method is in full rehearsal here.

René Descartes introduced the new form of skepticism, the foundational kind, in his *Discourse on Method.* His thought experiment is the

J. Hattiangadi, *Francis Bacon's Skeptical Recipes for New Knowledge*,
https://doi.org/10.1007/978-3-031-52585-8_10

famous Cartesian skeptical doubt. He challenges us, once at least in one's life, to doubt everything we may think to be true. We must imagine that we know nothing about the familiar objects we perceive around us, even about our bodies. We may even assume that everything we think we may know about the external world is false. But he slays this skepticism by arguing that even when I try to doubt everything, I know something indubitably: namely, that I am thinking (doubting). He used such a new form of hyperbolic doubt to establish self-knowledge. He introduced the famous indubitability dictum as a foundation of all knowledge: "I think (I doubt) therefore I am." From skeptical despair (the despair his new form of doubt created if left unchecked) he rescued an indisputable truth, a certainty. This proclamation is known fondly among philosophers as the *cogito* (short for its Latin form *sum cogito ergo sum*). For many people, not philosophers, this hyperbolic doubt may be all they know about the thought of René Descartes. For a philosopher, this thought experiment challenges all unfounded claims as falling short of knowledge. At least, compared to the certainty of his argument in the *cogito*, all other knowledge seems to fall short.

Foundational doubt is not obligatory for us to entertain. To engage in such doubt assumes that there is some well-founded knowledge. He suggested that we should engage in it at least once in one's life. If we do, we will see that only a valid foundation for all knowledge can pass his skeptical test. In this way, he shows or tries to show how one grand thought experiment is all we need to establish a foundation for all knowledge in science. In recent epistemology, philosophers often neglect to notice that the method of hyperbolic doubt, as this is called, is not forced upon us. Recognizing our failure to meet it, if that is what we find, is merely to admit that Galileo Galilei's method of thought experiment does not yield an indubitable foundation for knowledge. Such a result is somewhat trivial since we may find the knowledge elsewhere. If, however, we find one such argument, it provides us with a piece of knowledge as certain as any possessed by God, then it is a very important find. It is this that animates part IV of his *Discourse*.

In a certain sense, the *cogito* is the starting point of the Cartesian attempt to build a new understanding of the universe. It may be argued that knowing it is to know the spring that nourishes his thought at its very source. This interpretation of the *Discourse* would not be true. There is a little more philosophical background to this new philosophical beginning than meets the eye, even though the *cogito* was meant to be a new beginning for all knowledge.

In his *Discourse*, the *cogito* only appears in the fourth part of the six-part essay. It is neither the beginning nor the end but a salient point in between. The first three sections build up to the point where he begins building up a new science of nature; he is thus far into his thought by then. He intends to build a new structure based on his famous dictum, the *cogito*. However, long before he gets to his new, hyperbolic form of skepticism, he has slain the old ogre of ancient skepticism. The new form of doubt can only be entertained after he has overcome the ancient skepticism of Sextus Empiricus. We may add that he has also abandoned the possibility of Francis Bacon's new method of true induction. His foundational method is a rival proposal.

The *Discourse* is full of ingenious arguments, but it is also filled with a narrative account of the author's thoughts as a young student. The whole text is written up as an argumentative story. It is almost like a cinematic form of a narrative account of his thought, such is its gripping quality. His rhetorical brilliance does not take away from the ingenuity of his argument. His writing complements his arguments and philosophy. Suppose we interpret the *Discourse* as a moving picture with a well-defined plot. In that case, we must take note that the slaying of ancient skepticism takes place before the cinematic narrative begins—it is off-camera, as it were. The *Discourse* relates a story following that slaying. For a non-philosopher, this slaying may not be evident. Still, the mayhem is hard to miss for someone familiar with philosophy and knowing the contrast between ancient and modern skepticism.

René Descartes begins by asserting that all human beings have good sense, reason, or the ability to judge statements as true or false. In this, he says, we are all equally endowed as humans. We are unlike all other animals that lack this ability. Aristotle or Thomas Aquinas would have applauded this, but perhaps not what follows. Because we are rational, he said, we can judge statements as true or false. Therefore, he said, we must be able to evaluate some statements as true (and, presumably, some as false). The tacit thought here is that if we could not assess even a single judgment as true, then we would not have the ability to judge at all; that is, we would not be able to distinguish between true and false judgments. Note that the refutation of ancient skepticism lies in the simple fact that we must know something if we are rational. If we state this in contemporary jargon, we are language-wielding humans. We are rational beings. Therefore, we must know something or other. So, skepticism is false.

Pyrrhonian skeptics had elaborated upon a skeptical argument about the *criterion of knowledge* that they shared with the Academics. A skeptic may ask us how we can know what we seek when seeking knowledge. The second half of Meno's challenge to Socrates was just that. If we do not have a criterion to recognize what we identify, then there is no point in the search since we will not know that we have succeeded even if we come across it. But which criterion should we choose? Given any criterion, it has several competitors among other philosophical schools. Choosing a criterion for knowledge would seem to require a criterion of its own—a criterion for criteria. But this leads to the familiar dilemma that we must rely on circularity or infinite regress to begin the search. Neither option is satisfactory. A dragon has swallowed our method for seeking knowledge in the sciences even before we start our search. We note how Descartes slays that dragon.

Our reason saves us from ancient skepticism. Some ancient skeptics had asserted that we may not know anything to be true. Others had refrained from assenting to any nonevident knowledge. But we must know something to be assertably true, whatever that may be, notwithstanding the doubt of the ancient skeptics. We *must* be able to judge the truth or falsity of some statements to understand the language we possess. As humans, we have language and reason. Therefore, we can genuinely judge something. That judgment could be the foundation we seek for infallible knowledge of the sciences. It would be a kind of inner sense, a rational intuition. In a manner reminiscent of Plato in the *Meno*, Descartes suggests that our intuitive criterion of truth must be based on a simple recognition by our mind. Unlike Plato, he does not indicate a doctrine of recollection of Forms.

We note with interest that René Descartes slays all earlier forms of skepticism if we grant him his implicit argument. No longer can we assert that we know nothing. We must know all kinds of things merely by virtue of being rational animals. This argument, too, is reminiscent of

Augustine of Hippo against the Academics. Augustine abandoned skeptical philosophy because we know too much that the skeptic cannot allow. He opted for a belief that foregoes any foundations, a faith to which he gave his allegiance.

The Cartesian argument is unlike Augustine's and unlike Plato's, too, in distinct ways. Unlike Augustine's argument, René Descartes abandons skeptical doubt not for faith but for *logos* or reason—to find truth in the sciences. Unlike Plato, he does not make his criterion of knowledge depend upon recollection. Cartesian reason takes a recognizably modern turn. The modern theory is recognition without recollection. We must be able to judge truth and falsity naturally, by intuition. Such knowledge cannot be mistaken. It follows from our ability to reason, without which we could not have any human judgment or descriptive language. Without reason, we would not be characteristically human.

This argument marks the beginning of modern epistemology and its distinctive difference from ancient forms of it. The "rationalists" or "intuitionists" followed René Descartes in supposing that we can build all our knowledge based on our natural ability to judge some statements to be true - on "our intuition." The later empiricists in this tradition were critics of part of the Cartesian clan. Some philosophers agreed that we know some statements to be true as rational animals, but they denied that we can build a natural philosophy entirely on such intuitive judgments.

The argument that we must know something seems conclusive until it was shown to be fallacious by Bertrand Russell in *The Problems of Philosophy*. In 1912, Bertrand Russell distinguished between a *definition* of "true" and a *criterion* for truth. We need to have a coherent understanding of what the words "true" and "false" mean, i.e., an intuitive grasp of the definition of their truth (Russell 1912, 210–18). Such a definition underpins how we use statements in descriptive language. But we may not also have a *criterion* of truth for any statements. We do not need to know which descriptions in the descriptive language are true. Skeptics can deny we have a standard of truth, but they need not deny its definition. The Cartesian argument confused these two requirements. Their confusion lies at the heart of much of modern philosophy before Bertrand Russell made the distinction. It may also include some philosophy since 1912 propounded by those not paying attention.

The original empiricist objection to this Cartesian structure was as follows. Suppose we allow the knowledge of the content of sensations to be admitted. Our additional knowledge is severely limited. Merely inspecting a proposition may determine that it is true because of a relation between ideas within. These days, we would say that it is a truth based solely on the meanings of the terms used. We should allow knowledge only of "analytic" statements, in short. We can judge, "It is either raining or not raining." Knowing this tells us nothing about the world, for instance, whether it is raining. Any further judgment about the world would need to be assessed on perceptual grounds. We need to do more than examine the judgment all by itself. We see a lacuna between sensations and analytically true statements. Modern rationalists allow that there is intuitive knowledge of the world beyond analytic propositions.

The division of philosophers into rationalists and empiricists is a fundamental division of two schools in modern philosophy. Rationalists think that certain statements are known by merely inspecting them, though they factually describe the world. In contrast, empiricists deny that any information about the world can be known ultimately without checking it within the content of our senses. These are two different schools of modern philosophy. We should note, however, that both schools depend on René Descartes for having slain the ogre of ancient skeptical doubt, off camera, even before his narrative that leads us to his new method. Most modern philosophers are from two halves of one Cartesian clan.

The rest of the first part of the *Discourse* is a narrative concerning his dissatisfaction with the status of each branch of received knowledge that he had learned in school. This narrative is very nearly skeptical in its form. We can compare it to Michel de Montaigne's, who had decried all earlier knowledge claims or systems of philosophy. Francis Bacon complained about the Idols that beset Anticipations of the Mind. So, too, René Descartes writes about his dissatisfaction with each of the arts and the sciences in school.

He makes a singular exception, however, in the subject of mathematics. "I delighted most of all in mathematics because of the certainty and the evidence of its reasonings." This exception distinguishes his account of received knowledge from the usual skeptical disparagement. This exception confirms our suspicion that ancient forms of skepticism were indeed slain off-camera by René Descartes. As a young student in mathematics, he found examples of truths so clear as to be assertable. These examples were also distinct from one another so that we would not confuse a true for a false assertion. His method, he tells us, is designed to lead us to an extension of mathematical certainty to many more claims, to generate the true principles of natural philosophy—the true underlying principles of the universe in which we live. He was a foundationalist even as a young boy—if he is reporting accurately. And he was not faint of heart.

René Descartes complains that while he found some certainty of knowledge in mathematics, the subject had not been used previously as a basis for such a wider understanding of the universe. "… I did not yet notice its true use, and, thinking that it was of service merely to the mechanical arts, I was astonished that no one had built anything nobler upon its foundations, given that they were so solid and firm." (Descartes, AT VI.7–8). In "the mechanical arts," perhaps René Descartes included Archimedean thought. The noble structure he promises is a new way of founding a form of Galileo Galilei's mathematical physics. The new method of studying the geometry of motions was probably introduced to him by Isaac Beeckman not long after he had finished school. We cannot rule out the possibility that he was reading more into his earlier musings than was there.

Isaac Beeckman was likely aware of Galilean innovations from his University teachers, Willebrord Snell and Simon Stevin. It is this "physical mathematics" that René Descartes proposes to develop in his *Discourse on Method* as his new expansion of mathematics. He aimed to expand our knowledge of certain indubitable truths to new principles that will help us undergird natural philosophy. His natural philosophy would propound all the fundamental principles of the natural world. It would be based not on philosophical speculation, as in the past, but on a sure mathematical foundation, on clear and distinct judgments that constitute reasoned knowledge.

The new method of science that he proposed is designed to rely on only clearly true statements. These must be available to us as "clear"—i.e. as clearly, true or false. We know that we know some clear judgments because we have language and reason. So we must begin with those. He recommends avoiding all errors as we proceed from these clearly known truths by reasoning validly to those less known. Reasoning validly is judging truly what is before the mind. In a nutshell, his theory of the method of science is that if we begin with what is certainly known, and then if we avoid all error and allow our mind to be guided only by its natural ability to distinguish clear truths from falsehood, then we will have built a true new science-based entirely on our reason—on our natural ability to judge truly (Descartes, AT VI.31–40).

René Descartes had worked out a method in his unfinished *Rules for the Direction of the Mind* (Descartes, AT X.359–472). It was written in the 1620s, a decade before he had adopted Galileo Galilei's method of thought experiment after reading the *Dialogue*. The four rules set out in part II of the Discourse, and below, may be thought of as a gist of those pre-Galilean rules, i.e., before thought experiments came to occupy his thoughts on foundations.

When we judge those statements as true that are clear to us, we cannot be mistaken about them. René Descartes says mistakes can only arise concerning clear truths if we confuse two similar judgments. We find one to be clearly true and think confusedly that a similar but different statement is true because we have not *distinguished* them carefully. Whatever is clear enough to be in our foundation must also be distinct. The Cartesian criterion for the acceptability of a judgment as true is that it be "clear and distinct." The phrase recurs frequently in his writing.

He likens the project of pursuing knowledge to building a new house in which to live. He suggests that to develop our knowledge well, we must first tear down the old structure (which is full of errors) and begin with the indubitable truths. We must allow no mistakes to creep in as we build upon this secure foundation.

While we are building this foundation, how do we think and act? René Descartes suggests that since we act without the knowledge we have yet to build, we must act like Pyrrhonian skeptics. We do not yet know. We must accept the customs of the time and place as if they are binding on us. Once we have built our structure, we can also base our ethical knowledge on a secure foundation. After that, we will have no further need for our interim conservative lifestyle. This dichotomy between an ethics founded in true knowledge, and an interim ethics until enlightenment is achieved, is a theme that recurs in different guises in subsequent foundational philosophies.

He claimed that he had already used the following fourfold method with success: "The first was never to accept as true anything that I did not plainly know to be such; that is to say, carefully to avoid hasty judgments and prejudice; and to include nothing more in my judgments than what presented itself to the mind so clearly and so distinctly that I had no occasion to call it in doubt.

"The second was to divide each of the difficulties I would examine into as many parts as possible and as was required in order to better resolve them.

> The third this was to conduct my thoughts in an orderly fashion, by commencing with those objects that are the simplest and easiest to know, in order to ascend, little by little, as by degrees, to the knowledge of the most composite things, and by supposing an order even among those things that do not naturally precede one another.
>
> And the last was everywhere to make enumerations so complete and reviews so general as that I was assured of having omitted nothing. (Descartes, AT VI.18–19)

The first rule is often mistakenly used as a substitute for the Baconian method, especially given his Baconian use of the words "haste" and "bias." The second and third rules are his version of the ancient mathematical methods of analysis and synthesis, as found, for instance, in Pappus, *Collections Bk VII*. The fourth rule is his own and included to eliminate another source of error. It is added to ensure that we do not mistake a part of a solution for the whole of it. These four rules are well suited to solving mathematical problems, and rule two even mentions problems without specifying what kind.

To pursue his method of avoiding all errors, he had to find a beginning that was itself free of all possible errors. It must be indubitably true. Its basis had to be a judgment so clear and distinct that he could not even doubt it. Surprisingly, he did not find a new beginning in geometry or arithmetic, or even more generally, in a part of mathematics. Unlike the conjecture of Galileo Galilei, he does not find such nearly divine certainty in geometry or arithmetic. In the fourth part of the *Discourse*, he introduced the *cogito*. By doubting everything (this is the "hyperbolic" doubt), he found that there is one thing he cannot doubt: that he exists as a doubting or thinking being. He hoped to derive the most general understanding of the natural world from this indubitable, clear, and distinct beginning of self-knowledge. Here, at last, he had just slain the second ogre, the skepticism of hyperbolic doubt that he had just introduced into the philosophical literature as a new beginning for knowledge in the sciences.

II

Anyone unacquainted with René Descartes' accomplishments in mathematics, physics, physiology, and philosophy may think his foundational project a bit mad. At the very least, it seems too grandiose to be practical. Such is not the case. He was very successful as a natural philosopher. By the end of his life, he was perhaps known as one of the two of the most accomplished methodologists and natural philosophers of his time. Galileo Galilei was the other. Their thought dovetailed quite nicely on several key points and only diverged in some details. Consequently, the influence of his methodological thought on his contemporaries was enormous. This method works. That was his message in the *Discourse on Method*, which prefaces three brilliant essays in natural philosophy to showcase how well it works.

The influence exercised by his *Discourse* is such that even the third part, the least significant part of the *Discourse*, or so it seems at first reading, had a profound influence on contemporary and subsequent thought. In this part of the *Discourse*, he describes how he conducted himself as a person while he was building his new house. By the terms of his program, he could not claim to know anything until the new knowledge structure was built. To assume anything that is not well founded may introduce error beneath the new house, which cannot be corrected from above. But where should he live while the building is yet to be built? How does he conduct himself in the interim if he assumes no knowledge?

He adopts a recognizable Pyrrhonian strategy as his provisional moral code: he would live following his community's laws, customs, and religion. He would act resolutely even though his conduct would not be based on firm principles based on reason, i.e., on knowledge, but on a provisional code of conduct. He would act to improve himself rather than seek his fortune. Finally, he chose as his career the life of a philosopher inquiring into truth in the sciences, which is what the word "skeptic" means (an inquirer). Still, it is also appropriate enough for this brave slayer of skepticism until he has built his new structure of knowledge. The importance of this set of interim choices lies partly in temporarily adopting a very conservative Pyrrhonian strategy for morals. It also lies partly in the implied promise that the true moral code would be built upon the new foundation of ethical knowledge based on reason once the building is built. Then we would live by a code based on reason—not conservatively, not skeptically. His implied promise is that after a period of skeptical conservatism, we would graduate to a well-founded absolutism about how to live.

The Cartesian conception of an ethics based on reason inspired many rational and empiricicist philosophers. Benedict Spinoza, John Locke, Gottfried Leibniz, and others thought that ethics and politics must have a rational foundation. Immanuel Kant, a resourceful philosopher, eventually invented an astonishingly plausible ethics based on reason, which, despite failings, fulfilled the promise made by earlier modern philosophers, i.e., René Descartes and John Locke.

In the later part of the seventeenth century, philosophers sought rational ethics and politics. The time has been called the "Enlightenment" in Europe, from John Locke's promise of rational ethics and politics to its fulfillment in the eighteenth century by Immanuel Kant. Those with doubt that reason could properly generate such a great moral and political edifice, such as David Hume, and Edmund Burke, fell back on a more skeptical attitude to ethical conduct. The provisional Cartesian moral code became, for some, a permanent one. A non-foundational study of ethics and politics today requires a new initiative, but thinking about it is beyond the scope of this essay. We should note at least that the parameters and limits of the modern study of ethics, between reason and skepticism, are laid down in the third part of *Discourse on Method*. René Descartes, however, had little more to say on the substance of ethics and politics.

III

Self-knowledge is minimal if we cannot know anything beyond the claim, *cogito ergo sum*, "I think therefore I am." René Descartes suggested that the idea of God that he possessed within him showed him that God exists. The idea of God is a being who is unlimited, whereas I cannot think of myself as a perfect being, says René Descartes, because I am limited. Therefore, it seemed to him there must be an unlimited and perfect being on whom his existence must depend. From God's existence, he then goes on to suggest that since God cannot be a deceiver, we, his creations, can trust our senses to some extent (to the extent that it conforms to the dictates of reason) as a sign of things in the external, physical universe.

René Descartes employs the method of thought experiment again when he works out his metaphysics or the list of fundamental substances in the world. Aristotle listed all individual particulars as substances, or primary beings because they can support contrary attributes. These attributes, in turn, cannot each support contrary attributes. René Descartes introduces his metaphysical scheme for substance. He claims three primary substances: The self (selves), God, and matter. In part IV of his *Discourse*, René Descartes makes the case for the self as a thinking substance. The argument is that its essence is thought since it is indubitable that I exist as a thinking self, but I could doubt everything else about me. God is part of this argument because it is from my imperfection that I infer the existence of God. So it is inconceivable that God is not perfect in every way. It is unthinkable as the ground of his existence that God may lack any positive or good attributes. It follows that God's essence must be infinity or the lack of limitations.

That matter exists is not provided by a thought experimental proof, but its essence is worked out in another thought experiment unrelated to the *cogito*. Here is a summary of it. Take a ball of wax and notice that it may be solid and hard, but it will turn soft or molten or liquid when heated and vaporizes when heated. Even more, I can think of it as now having many qualities, but one thing I cannot imagine it not having is an extension. It cannot lack length or breadth of depth and remain matter.

Therefore, matter's essence is its extension (or its spatial extent.) Apart from the last sentence, this thought may have come from Galileo Galilei's discussion of matter in *Il Saggiatore*.

The definition of matter as an extended substance has many important consequences for the physics and physiology following it. It implies that there is no such thing as a vacuum. All locomotion must be cyclic, since bodies have no empty spaces to move. One body acts on another by pushing it out of the way. That body, in turn, must push another away. Eventually, some matter must come back around to occupy the spot just vacated by the first body in the hidden cycle. In part five of the *Discourse*, we find a description of the solar system based on his "rational mechanics." His claim was described at length in his suppressed book *The World*. But he had to make no reference to the moving earth, which was disallowed. There is also an entirely novel mechanical model of human (and all animal) bodies, which also features Harvey's discovery of blood circulating in the body. The animal as a machine is one of the consequences of the mechanism he advocates. In the *Optics* we find his discussion of the physiology of vision to be fully mechanical concerning how images reach the brain. The account of vision remains mechanistic until the translated image, or, more accurately, the nervous signal, reaches the mind.

Some of his contemporaries severely criticized the argument for the existence of God and the idea that God is a guarantor of some knowledge. Some argued that if God must guarantee any knowledge, particularly any rational knowledge, then that is part of a viciously circular argument: first, they said, you deduce the existence of God, relying on reason, and then argue that God guarantees reason. Circular argument!

In the *Meditations on First Philosophy*, René Descartes takes up the criticism by denying that he was ever arguing in a vicious circle. The fourth part of the *Discourse on Method* is greatly elaborated upon in the ten parts of *Meditations*. René Descartes does not argue that we can rely on reason because God warrants it. His argument is subtler. He argues rather that any knowledge derived from our mental states cannot be relied upon, whether these states be perceptual states or memory. To the extent that a rational argument is long, and we need to rely on remembering its earlier steps at a later stage, our mental states could betray us. We can rely on them to some extent. When we reflect that since God is not a deceiver, we may conclude that they are perhaps not deceiving us entirely—though they may not be transparently veridical, either. These states give us defensible knowledge only if we deal with them appropriately.

The case of memory is critical here. The Cartesian conception of a valid proof is one where the mind directly sees the truth of the statement proved. Reason is a form of direct intuition of truth. The criterion of "clear and distinct" ideas implies it. Arguments, however, can be long and tortuous, where we show that if *a* is true, then *b* is true, and if *b* is true, then so is *c*, and so on to a distant conclusion. Each step of this proof may be clear and distinct. Still, we must rely on our memory when remembering the first step just as we try to clarify a later step. At that last stage in the proof, we may be unwilling to trust our memory about our previous steps. If we could see all the steps in our mind at once, in a flash, we would know the truth to be proved, for then it would be based on reason. Such was the Cartesian standard of proof. As long as we rely on memory, proof becomes dubitable.

Consider a person who goes through a great many mathematical steps of an old proof. Eventually, she puts all the steps together into one giant step in which the truth to be proved is seen to be true. Does she know the truth at the very end? The full proof is before her mind at that stage, so she can truly judge. But until she reaches that stage, she depends on memory to reach that stage. Hence, even though reason does not depend on God's guarantee, "reasoning," or the process of recognizing a long proof, may force us to rely upon memory and therefore perforce God's goodness to see us through.

There is a certain sense in which the reliance on reason in Cartesian thought is ultimate. It remains unquestioned and unfounded. Of course, he thought God made us rational beings. But this was not offered as proof of our reason. Unsurprisingly, René Descartes was very concerned to show no vicious circularity in his argument. He was so concerned that he invented a novel way of showing that he was not assuming the existence of a benevolent God to prove his existence via the *cogito*. Once introduced, this novel idea had a potent impact on subsequent philosophy. It generated a new philosophical movement. Therefore, if for no other reason, we must pay attention to it. It was another thought experiment to prove that he was not arguing in a circle concerning God as a guarantor of our knowledge.

Far from assuming the existence of a perfect God to prove His existence, which would be viciously circular, René Descartes claimed that he could prove whatever is needed by assuming the contrary. He argues that the claim that God does not deceive us can be established beyond doubt if we begin by considering, contrarily, that there is an evil and all-powerful daemon who is deliberately trying to deceive us. This imaginary evil daemon is granted such powers for the sake of argument that he can make me think thoughts by putting them before my mind. I may think I am standing before you, fully dressed and speaking to you, even though I am at home beside my hearth in my dressing gown. We can give such power to this imaginary daemon that if I think I have hands, feet, or a body, it may be because he has put that thought in my mind. The evil daemon may have placed anything before my mind, even if faithful to reality.

I may use the "evil daemon hypothesis," as it has been called, to doubt that my body, my hands, my feet, and the entire world about me is real, the one thing that I still cannot doubt while I am doubting, is that I doubt. I think, therefore, I am. My existence cannot be doubted even based on a special assumption of an evil daemon, a skillful deceiver. Moreover, once the *cogito* is allowed, the rest of his argument goes through just as if we had not assumed the evil daemon. The Cartesian account of an imaginary evil daemon was proposed as a rhetorical device to show that even if we make that assumption, the Cartesian proof goes through. René Descartes may claim, therefore, that his argument does not presuppose the benevolence of God. Subsequently, we can argue that God is not a deceiver. That is his thought experimental argument. But it has other important consequences for modern philosophy.

Even if we come to doubt everything about the external world, there are two things that even the evil daemon cannot take away from us, René Descartes claimed (*Meditations 3*): 1. I exist as a thinking thing, and God exists, and everything else that may follow from those pieces of knowledge; and 2. even when my senses are being deceived by an evil daemon who puts those ideas in my mind, I cannot be deceived that those very ideas are present before my mind (even if they do not depict the natural world as it is).

We may rely on two kinds of knowledge to build our new science. There is that pure and austere rational knowledge, such as that of the *cogito*, the existence of self and God, and the fundamental principles of mathematics derivable from it, as they apply to the natural universe. There is also experience: confused and rich but not entirely trustworthy knowledge we obtain from our senses. We can succinctly state the Cartesian method for finding truth in science: It is to seek out and then rely on the pure, austere principles of philosophy to guide us in separating truth from error in our rich information from the senses. The basic principles of natural philosophy, which we establish first, guide us to a future state of knowledge so detailed that we can then use it for our benefit in practical arts, such as medicine.

Only in the sixth and last part of the *Discourse* does the full scope of the method for discovering knowledge in the sciences become clear.

IV

He does not list them for fear of censure, but he lets it be known that he had discovered the fundamental principles of philosophy by his new method. He had also worked out the most common and simple effects of these causes. But *Discourse* part VI finds him making this remarkable admission, "Then when I sought to descend to those effects which were more particular, I encountered such a diversity of them that I did not think that the human mind could distinguish the forms and species that are now on earth from that infinity of others that God could have put there if it had been God's will to put them there…." The principles of philosophy cannot tell us which things exist. We need our senses to tell us this. What will the abstract Principles do to make sense of these objects that we perceive as opposed to those that might have been but are not?

"Consequently, I thought that the only way to making these objects useful to us was to progress to the causes by way of the effects and to make use of many special observations…." (Descartes, AT VI.64). This means that we must go from the senses back to theories about the objects, by way of special observations, whatever those are.

> ...And now, reviewing in my mind all the objects that have ever been present to my senses, I venture to say that I have never seen in them anything that I could not explain quite easily by the principles that I had discovered. But I must also admit that the power of nature is so ample and vast, and these principles so simple and so general, that I notice hardly any particular effect that I do not know at once that it can be deduced from the principles in many different ways; my greatest difficulty is usually to discover in which of the ways it depends on them.... (Descartes, AT VI.64–65)

In short, we must assume some hypothetical models of an object to understand what we find by our senses, and his greatest difficulty is in deciding which hypothetical model works best, given the principles he has discovered.

He continues, "...I know of no other means to discover this than by seeking further observations whose outcomes vary according to which of these ways provides the correct explanation." When he has a model, he must look at its effects to see if it fits further "special" observations. These effects will refute the poor models and confirm the good models as explanations of the object to be understood. But the process is indeterminately long.

The Cartesian method allows us to discover the true principles of philosophy: the laws of motion. Of course, this is a controversial Cartesian claim. Having found them, we can try to understand objects by seeing how the first principles apply to them. Since our sensory information is very detailed and diverse, we find the first principles in the objects of the senses employing a hypothetical model (consistent with first principles) that we use to derive its features, and we need to make a special observation to refute the models that are not correct as explanations.

We can observe how this method works in practice because René Descartes gives examples of its deployment in two of the three essays that the *Discourse* prefaces. But one thing is glaringly obvious. If there are too many hypothetical models from which to choose, then we have little more grasp of reality than the knowledge of first principles.

We have our abstract knowledge of the self, God, and matter. What else can we know of the external world based on intuition alone? Suppose we adopt the empiricist rule that nothing concerning natural philosophy can be discovered except through the senses ("no armchair physics," as some physicists admonish). In that case, intuition alone does not know the first principles of nature. We must reject Cartesian first principles, too. As empiricists, we are left with only the second half of the Cartesian method to deploy, as described in part six of the *Discourse on Method*. The hypothetico-deductive method, as it is called today, is all we have left for the method of natural philosophy.

V

This book explores Francis Bacon's method, which is poorly understood today. To fully appreciate it, we must study it in an application, and its first successful application is in Isaac Newton's optics. He employed the experimental philosophy modified in the Royal Society of London from Francis Bacon's original expression. "Natures," "Ideas," and Forms" were abundant in Baconian writing. These concepts ran against the current of mid-seventeenth-century thought. That thought had become heavily influenced by Galileo Galilei's and René Descartes' new mathematical and mechanical theories of natural philosophy. The experimental philosophy was brought up to date by Robert Boyle. He modified the Baconian method by adopting mechanics as a hypothesis. He also adopted the corpuscular hypothesis without committing to either atomism or plenism. The new mechanical and mathematical philosophy had also arrived with a new foundational method associated with it. Not all philosophers adopted the new method. Those who admired mechanics but not the method of the *Discourse* included the early Fellows of the Royal Society of London. There are good reasons for this. They adopted Baconian induction partly because the most promising Cartesian method soon lost its luster.

Galileo Galilei's thought, its forerunner, was widely admired among mathematicians and philosophers, but it was not a complete system of philosophy. It was very suggestive. Many of his suggestions had not been discussed enough in detail to be adopted. René Descartes, on the other hand, proposed a comparatively complete mechanical philosophy, much more systematically worked out. Many hoped to replace the schoolmen's philosophy with the new Cartesian mechanical philosophy. However, some would not accept the method that seemed naturally associated with it. Their reluctance was because of two difficulties that arose for the Cartesian method.

René Descartes claimed that the laws of motion propounded in the *Principles of Philosophy* are clear and distinct and well-founded on the *cogito*. Far from that, they turned out to be false because they were found logically inconsistent. Christiaan Huygens discovered, to his great chagrin, that René Descartes' laws of impact made essential use of the difference between bodies at rest and those in motion. In contrast, the law of inertia implies that motion and rest are equivalent. The laws of impact violated the inertial law!

It can shake one's faith in a method if it promises certainty of truth but leads us to error and falsehood. To evade the conclusion, one can always defend a method as excellent by blaming its flawed application. Christiaan Huygens and most Cartesians on the Continent of Europe adopted this approach by improving the laws of impact while requiring rational, mechanical, explanations of all phenomena.[1] But the Cartesian method also had a second difficulty, which we noted earlier. Inconclusive modeling was a limitation of the method itself. Even if we use it best, we must assume hypothetical models to apply general principles to particular things. One could use mathematical probabilities to bridge the gap by showing some hypothetical models to be superior. Christiaan Huygens, Blaise Pascal, and Pierre de Fermat developed a mathematical account of probability based on finite gaming models. They hoped for a theory of probability that would generate knowledge of the best hypothesis—knowledge that is one step removed from things. We can also avoid hypotheses in another way. Using the Baconian inductive method to generate models of particulars, we may replace hypothetical models with knowledge of real schematisms.

Fellows of the Royal Society of London wanted knowledge, not suppositions. Under Robert Boyle's influence, the Society adopted the updated experimental philosophy. Fortunately for Isaac Newton, applying the updated version of the Baconian method allowed him to make remarkable contributions to natural philosophy. In optics, he discovered that sunlight is only apparently white—it conceals a mixture of colors.

Cartesian optics was much admired at the time. Isaac Newton studied it in depth under the guidance of Isaac Barrow. When graduating from Trinity College, he took up optical theory, where René Descartes had left off. We must first study René Descartes' brilliant application of his foundational and hypothetical method and then compare it with Isaac Newton's application of experimental philosophy. Thus, we can better appreciate the difference between the two competing methods of modern science: foundational (hypothetical) and skeptical (inductive).

Note

1. Christiaan Huygens was a lifelong Cartesian, though he had reformed it by abandoning the conservation of quantity of motion. He had the greatest influence on Isaac Newton, who studied with great carfe the former's book *Horologium oscillatorium dive de motu pendulorum ad horologia aptato demonstrationes geometricae* (1673). In that book, Christiaan Huygens took up Galileo Galilei's study of the pendulum for further mathematical analysis. He reformed the Cartesian law of impact, asserting instead the conservation of momentum (not motion), which is equivalent to Isaac Newton's third law, that action and reaction are equal and opposite.

Reference

Russell, Bertrand. 1912. *The Problems of Philosophy*. London: Williams and Norgate.

11

Applying the Foundational Method

I

The Cartesian method of science is founded on demonstrative knowledge, namely, the rational first principles of motion that he had derived from an indubitably true foundation. They were to be the pride of his *Principles* a few years later. But he could not divulge them in his 1637 publication. Discussing principles of locomotion took him too close to what the Church had forbidden. A few years later, he found a way to satisfy the Censors in his *Principles of Philosophy*, but evidently, that clever maneuver had not yet occurred to him in 1637. Using his new method, he used it in two of the treatises to illuminate some optical phenomena. These include the famous new *sine* law of refraction. He then combines these results with hypothetical models invented to fit our experience of the colors of rainbows. His explanation is impressive enough for Isaac Newton to remember it as the "celebrated Phenomena of Colours." (Newton 1672, 3075).

Isaac Newton's method of experimental philosophy is very different. He does not begin with the certainty of first principles. He starts investigating a discrepancy in the appearance of a refracted image. Then he

J. Hattiangadi, *Francis Bacon's Skeptical Recipes for New Knowledge*,
https://doi.org/10.1007/978-3-031-52585-8_11

collects more odd phenomena to make up a short but adequate natural history that shows him sunlight has hidden within it mixed rays of colored light. He provides a unique solution to the discrepancies collected for the purpose. He then proceeds to deploy this model to make sense of several other puzzling color phenomena that Robert Hooke and Robert Boyle had reported earlier. The contrast between the two methods is instructive. Each is brilliant at using his method—one using hypothetical models to explain some phenomena mathematically and the other working out many evidentiary discrepancies to extract a mathematical model from the puzzling facts themselves.

II

René Descartes' treatise, *Optics* (*Dioptrique*), begins by telling us that he will not tell us the true nature of light. Instead, he made use of three comparisons. They "… help to conceive it in a manner which to me seems the most convenient to explain all those of its properties that experience acquaints us with, and to deduce afterwards all the others which cannot be so easily observed; imitating in this the Astronomers, who, although their assumptions are almost all false or uncertain, nevertheless, because these observations refer to different observations which they have made, never cease to draw many very true and well-assured conclusions from them." (Descartes, AT VI.83).[1] He explicitly invokes the famous "method of hypothesis" to "save the phenomena." Plato's *Timaeus* introduced those ideas to astronomers. Then René Descartes proceeds to give us his three comparisons for light. He compares seeing eyes to a blind man feeling objects around him with a walking stick. The medium of light is also likened to a partially filled vat of crushed grapes in which wine is made. And when light bends at an interface between two media, he compares light to a tennis ball being hit onto a surface. His method is rich in suggestive models. His results are astounding. These models exhibit René Descartes' method at its impressive best.

A walking stick in the hand of a blind man is an excellent comparison of light as René Descartes conceived it. He believes that extension is the essence of matter. Since any piece of matter is whatever it is with geometric precision, he envisions it as perfectly rigid. It cannot become more or less than it is. Otherwise, two unequal lengths would equal, which makes no mathematical sense. Light must be transmitted instantaneously in the intervening medium since the medium is perfectly rigid. It does, however, have a pace or a speed. "I would have you consider light as nothing else, in bodies called luminous, than a certain movement or action, very rapid and very lively, which passes towards our eyes through air and other transparent bodies, in the same manner that the movement or resistance that this blind man encounters is transmitted to his hands through the medium of his stick." (Descartes, AT VI.84)

We must note, however, that the medium transmits pressure in a straight line. Pressure is a "tendency to motion." It is not a motion. René Descartes decrees that light pressure obeys the same laws as motions. He compares light to the pressure built up in a vat of wine. Imagine a vat with crushed grapes, with two holes stoppered at the bottom. If the vats are opened at the stops, the liquids will start flowing out of the opened holes. But if kept stopped, the wine presses down on the stoppers, with a tendency to move out without actually moving. We would observe a rate of movement if we uncorked the holes, of course, but when stopped, the wine merely exerts a certain amount of pressure in proportion to the movement of the wine were it unstopped. Take three points, C, D, and E, in the middle of the vat of wine. Name the two stoppers at the bottom A and B. Pressure is then exerted in straight lines from each of C, D, and E directly to A and a straight line from each of C, D, and E to B. Similarly, we are to conceive of the light from a luminous body as a pressure traveling in a straight line instantaneously towards each of many objects and to our eyes simultaneously.

There are semi-solid or crushed grapes in the vat, but there is also the finer kind of liquid, or wine, filling the spaces between them. We may imagine that even a subtler pervasive matter called "ether" fills any spaces between particles of wine so that indefinitely many lines in every direction carry the pressure of light everywhere from a luminous body through this subtlest of all media, the ether.

Moreover, we should note that though a velocity is associated with pressure, the pressure is instantaneous. If a stopper is released, the wine has a velocity gushing out of a hole. But when a plug or stopper blocks the hole, the transmission of the pressure in wine is static but instantaneous, according to this account. It is transmitted instantaneously in a perfectly rigid medium (space or matter). Therefore, light *as a pressure* travels instantaneously in the medium of the ether. The light from the sun takes no time to reach the eye because it is a pressure or a tendency to move and not a motion. The tendency to move has an underlying rate of movement though it is static. The rate determines not how fast the pressure moves (it is static) but how great the pressure is.

Nevertheless, the pressure or the tendency to move obeys all the same laws as motions. So as a third comparison, René Descartes asks us to imagine that light is like a tennis ball that moves when struck by a racquet. Since we can imagine this more readily than pressure, we can use that comparison to study the observed properties of light. We can visualize what happens when light, like a tennis ball, hits and reflects off bodies or as it is bent or refracts as it passes through from one transparent medium to another. We assume a general rule that in any one medium, light travels in a straight line. When light encounters a second medium, then it either reflects or refracts. To understand optical reflection and refraction, we should first study the model of a tennis ball hitting different surfaces. It is much easier to imagine motions and then transfer this intuition to the instantaneous action of varying pressures in other media.

We should ask ourselves before proceeding any further: how can this be a proper method? We have been given three completely different models of light. The three models of light presented to us cannot be true at once, so they cannot be used together. He thinks of each of the comparisons like diagrams a geometer offers. No diagram is essential to proof. It is included, he would suggest, just as an aid to the student's imagination. The proof itself needs no diagram. A student may need them to follow the proof, perhaps. Our job is to see how mathematical generalities illuminate the particular.

Light travels in a straight line in a transparent medium, as we can ascertain by experiment. However, when light encounters a new medium, it may be entirely reflected whenever the second medium is opaque. It may also be deflected by a slight change in angle (refracted) when the second medium is transparent. We understand reflection and refraction by appreciating the underlying principles of motion.

René Descartes uses two mechanical principles that he shows underlie both reflection and refraction: one is the conservation of the quantity of motion; and the other is the mutual independence of two rectangular components of any motion.

The conservation of the quantity of motion was like the conservation of the quantity of matter. We are familiar with the latter idea: God put this much matter into the Universe. This quantity of matter cannot be increased or decreased by bodily interactions. In the same way, René Descartes could say that God put so much motion into the world. It can be redistributed through interactions, but motion cannot be created or destroyed. Modern physics does not treat motion in this way. Cartesian physics asserted the conservation of the quantity of motion.

The independence of rectangular components of motion is an old principle. It was known in statics since ancient times. However, it became widely appreciated only in the seventeenth century, after Galileo Galilei used it in his new kinematics, his physical geometry of trajectories. A motion at any point may be decomposed into two component motions at right angles to each other at that point. We can use a force to oppose the component of motion in one direction. The other component, being at right angles, will remain unaffected by the applied force. The resultant motion will combine one component unchanged while the other is modified proportionately to the force acting on it.

Using these two principles, René Descartes demonstrates two laws of optics. He demonstrates the law of reflection, known since ancient times, that the angles of incidence and reflection are equal. He also propounds and proves his newly discovered law of refraction when the light goes from one transparent medium into another transparent medium. The law he demonstrates is that the *sine* of the angle of incidence is proportional to the *sine* of the angle of refraction. Discovering this relation was the coveted prize of ancient geometrical optics. Still, it eluded Euclid, Claudius Ptolemy, Ibn Al Haytham, and Johannes Kepler. René Descartes' proved it with great ease by employing his method.

He speaks first about the reflection of light "to make the knowledge of refraction easier." In both cases, the model he uses is that of the tennis ball, moving at a constant speed from A towards B on the hard flat ground CBE where it is reflected away from the ground in some new direction

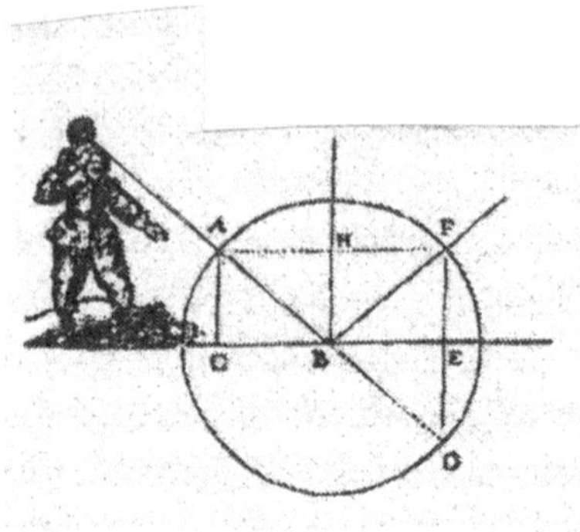

Fig. 11.1 Reflection

(see Fig. 11.1). We have to find that direction. That direction, which we need to find, is part of what René Descartes calls its "determination." Finding the path provides the law that tells us that the angle of reflection is equal to the angle of incidence.

René Descartes asks us to note that when the ball arrives from A to B, the "force of its motion," which is proportional to its total quantity of motion, or its bulk multiplied by its speed, is unchanged by the collision with flat ground CBE at B. The determination of the movement must change at B if the ball cannot go past the ground on its initial determination (See Fig. 11.1). Let us draw a circle around B with AB as the radius.

Since the quantity of motion must remain unchanged, the ball must end up on the circle's circumference again to travel at the same rate as going from A to B.

René Descartes now asks us to note that when the ball hits the surface at B, its motion may be compounded from two lines at right angles to one another; one is the perpendicular at B, being BN (the "normal" at B); and the other parallel to CB being the line AN which is perpendicular to the normal BN. The perpendicular component of the motion at the ground at B (given by BN) is wholly resisted by the ground CBE, whereas the component of the motion which is parallel to the ground, represented by the line AN, is wholly unaffected by the bounce off the ground at B.

The unimpeded parallel component of the motion AB is AN. It continues unchanged.

1. The ball must end up on the line DF parallel to BN and the same distance as AN at each point.

There are only two points where the line DF and the circle intersect and only one, at F, where the ball can end up if it cannot penetrate the ground. The ball must move towards F.

The two right triangles, ABN and FBN, have the three corresponding sides equal and are, therefore, congruent. Therefore, the angle of incidence, ABN and the angle of reflection NBF are equal, proving the law of reflection.

It is useful to go through this carefully because it makes the law of refraction easier to follow.

Consider Fig. 11.2. The tennis ball arrives from A to B and is deflected in a new direction at B because it is met by a flimsy cloth that breaks upon impact. Now we suppose that some of the force of the ball's motion is lost or transferred to the fabric.

Suppose the ball loses half of its motion to the cloth when it breaks through. Then it will take twice as long to travel the same distance after impact as from A to B before. Draw A circle around B. This time the line EF to which the ball travels to the right must be twice as far from B as the line AC is from BN. The ball must travel to the circumference and a

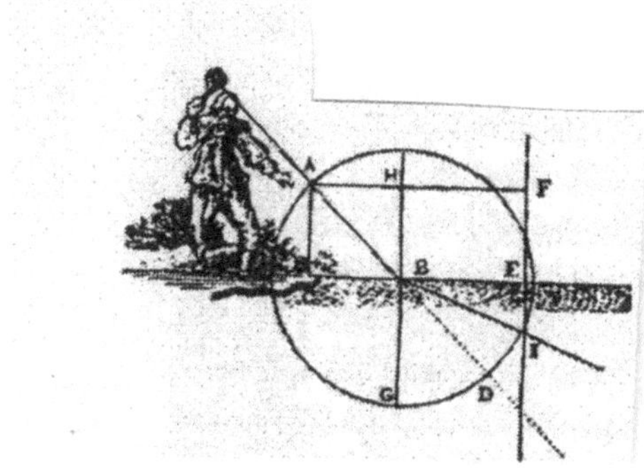

Fig. 11.2 Refraction

point on the line. Since there are only two such points, and the ball must go down (not up) from B to be refraction, it must arrive at the point I. Let us describe that angle which is the direction to I. Though some contemporaries of René Descartes did not appreciate this fact, the *sine* law follows from the two principles of motion alone.

Let us summarize the two principles we have used; one is that by a certain factor, n, multiply the quantity of motion.

(1) $Vr = nVi$,

where *Vr* is the quantity of motion after refraction, *Vi* before incidence and *n* is the constant of proportionality.

(2) $Vi \sin i = Vr \sin r$

since the ratio of the side opposite the angle represents the parallel component and the hypotenuse represents the total quantity of motion, and this ratio does not change.

Now if we substitute *nVi* for *Vr* and then divide both sides of the equation by *Vi* we get

(3) $\sin i = n \sin r$.

The constant *n* introduced by Descartes in his model to account for the change in speed is now the index of refraction going from one medium to the other. Going back from the second medium to the first, as light can be directed, it becomes i/n. René Descartes has given us mathematical proof of the law of refraction, which is remarkably free of the details of any refractive model. We use the model of the tennis ball as an aid to the mind, which the proof does not require. This proof shows his physics at its best. Isaac Newton's notes in his optical lectures mention that this proof is not inelegant (Sabra 1967).

Another case taken up by René Descartes is of interest. If we consider light traveling at a very sharp angle to a surface (see Fig. 11.3), then the time it takes to travel to the right after B takes it beyond the circle drawn around B. Now the line cannot meet the circle at all. Refraction fails to

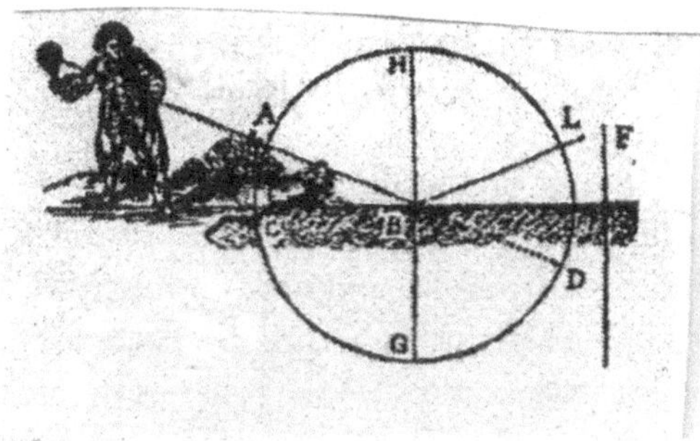

Fig. 11.3 Total Internal Reflection

occur in this case, and we get a "total internal reflection." In this case, the light bounces off or reflects because there is no room to refract. It is at too sharp an angle.

We arrived at these analyses by considering a tennis ball moving at a constant speed and penetrating through a barrier (or not). We have supposed that the model is unnecessary, but it was interposed in the argument. Light is a kind of pressure that "obeys the same laws as motion." Therefore, the same analysis must apply to pressure, he suggests. Like the tendency to motion in a perfectly rigid walking stick, the pressure travels through the ether instantaneously, taking no time. His analysis must apply temporal notions to its path, though it sounds paradoxical.

Pierre Fermat thought this shows that the proof was not mathematically adequate because its assumptions were based on analogies riddled with difficulty. He devised another proof of the *sine* law using the principle of least time. That was an important new event, but we do not need to suppose with Pierre Fermat that the Cartesian method has failed. As we have seen from the above reconstruction of the derivation, it did not require the models to be valid as part of the proof. On this point, René Descartes was right concerning his proof of the *sine* law of refraction.

René Descartes also notes, "Perhaps in doing these experiments, you will be astonished to find that the rays of light are more sharply inclined in the air than in the water, on the surfaces where the fracture occurs, and still more so in water than in glass—quite the opposite of a ball which is inclined more sharply in water than in the air, and which cannot pass through the glass at all." He notes a discrepancy, in short, between the motion of a ball and a ray of light in the water. But there was a ready explanation forthcoming.

He asks us to think about the model of a vat of wine, in which the pressure from its weight is transmitted to the bottom through the fine liquid and not primarily through the grapes. In much the same way, light is a pressure that transmits through the ether that permeates air and water. The paradox is dissipated if we take ether as the medium for transmitting pressure. This maneuver, too, is typical of the Cartesian method, in which every difficulty is dealt with at once. The resulting understanding is lucid at each step.

In the air, which is less dense, the ether is denser because it takes up all the free space. In water, denser than air, ether is less dense because it has less space to fill. The density of a material and the ether in it are thus in an inverse ratio. Therefore, the light behaves as it is to be expected in its proper medium, namely, in the ether, if it moves as a ball does in air and water.

III

Many readers of these models thought them inadequate. René Descartes provided an adequate explanation of this optical phenomenon. They were not proofs but hypotheses. Here he was using his models more substantively, but they were not satisfactory to ground proof. Christiaan Huygens, for instance, was not satisfied and later gave another derivation of the *sine* law based on the wave theory of light. Moreover, the principle that the force of an object in motion is given by (or is in proportional to) its quantity of motion, (that is to say, a quantity given by its bulk times its speed) was also given up by Christiaan Huygens. He substituted a different law, a law of the conservation of what we would call momentum, or mass times velocity, or the quantity of directed motion. Isaac Newton took up Christiaan Huygens' reformulation in his physics as correct. They were both reformed Cartesians in this respect.

One consequence of the confusing models of light proposed is that his contemporaries denied René Descartes authorship of the *sine* law of refraction. They said it could not have been his discovery or proof because his thought was too confused. In an unpublished manuscript written by Willebrord Snell, which has since been lost, there was an equivalently stated *sine* law of refraction. Several people, including Christiaan Huygens, saw the *sine* law formulated in that manuscript as early as in 1632. The *Optics* (*Diotrique*) was published with the *Discourse on Method* in 1637.

For this reason, the *sine* law of refraction is now known as Snell's law. It is sometimes even assumed that René Descartes found it there and came up with a spurious proof to claim it as his own. If we go by unpublished sources, however, this law was known to René Descartes before 1627, when he asked his friend Claude Mydorge by letter to order some lenses to test his discovery of what seems to be in context an older version of the *sine* law. It would seem that even if he had seen Willebrord Snell's manuscript in or around 1632, the *sine* law was old news to him. In any case, it seems uncharitable to assume that he did not independently make the discovery, even if it had been independently discovered even earlier by Willbrord Snell. The Cartesian proof is so lovely that it is hard to deny him some credit. The critical point here is that we must recognize in his proof what represents the *sine*: the independence of the parallel component of the motion of light. That is the main discovery here. The ratio of that parallel motion to the total motion gives us the law, i.e., that the two angles are proportionate by their *sines*. It is so simple and elegant.

These features of the thought of René Descartes are less relevant to our purpose in studying the application of method than they would be for an account of the history of natural philosophy. Our concern is only to see how the method was applied. In the case of the *sine* law of refraction, the proffered hypothetical models (pressures, spherical things) can be confusing, especially when run together. The principles of motion are very clearly being deployed, even if, like Christiaan Huygens, we do not agree with one of these principles. Their use illustrates how his mathematical method of hypothesis could illuminate aspects of nature that had eluded ancient optical theory.

IV

René Descartes also used his method to make sense of experience, as he does with the refraction of light from air to water or vice versa. This example is in line with his remark about the limitation of his method in the sixth discourse, which brings up a hypothetical model or extension to bolster his method. A brilliant example of this extension is his use of his optical models to shed light on the phenomena of the rainbow in the eighth discourse of his *Meteorology* (Descartes, AT VI.325-44).

Having argued in the earlier discourses of *Meteorology* that clouds consist of water droplets, he proceeds to discuss the rainbow in the eighth discourse as an extension of his optical theories of reflection and refraction. He makes a larger model of a drop of water, a large clear spherical glass flask filled with water. He notes the angles that the sun, as the light source, makes to the eye at the glass globe when he sees colors in the glass. (see Fig. 11.4) He moves his large model droplet around to record the angles at which the various colors appear. In the sixteenth century, the angles at which double rainbows appear in the sky were measured, i.e., the angle made by the sun's rays when deflected to our eyes at the rainbow. A bright rainbow was reported at an angle of about 45 degrees, and an inverted faint rainbow at about 55 degrees. These angles were experimental and approximate, of course. The light refracted through the glass

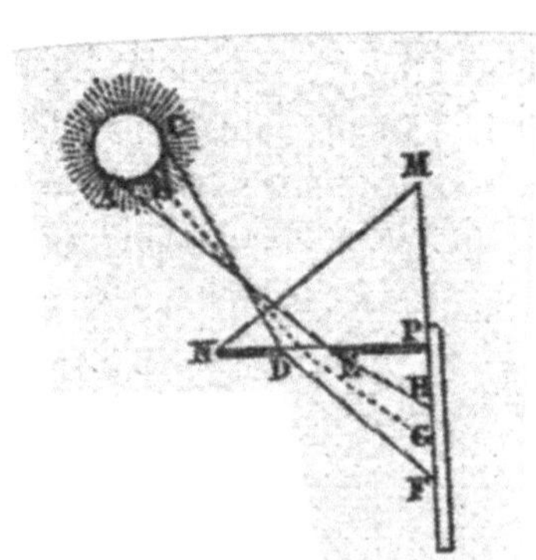

Fig. 11.4 Spectrum

model showed the colors at the same angles, confirmed approximately on René Descartes' large glass "droplet of water."

René Descartes shows his method and optical theory to good advantage by explaining the puzzling phenomena of the rainbow. First, why do we see that array of colors from violet and blue to yellow, orange, and red? Second, why do we observe them at those very angles that we do and not at other angles? And why in that same order and none other? Third, why are there always two such rainbows? Fourth, why is one always the inverted image of the other? Fifth, why is one of the inverted images always fainter than the other? All these questions are answered in the text of the eighth discourse, though the text does not pose these questions in this manner.

To explain colors, he takes a prism and sends sunlight through it (see Fig. 11.4). As the diagram shows, the prism is held with the face perpendicular to the sunlight. The sunlight refracted through the lower or farther face of the prism is refracted with colors in it. We see a full spectrum of colors going from violet to red on the screen held beyond the prism. We note that the light is at an acute angle to the surface of the prism as it reenters the air below. Moreover, we note that the angle of refraction is different at different ends of the beam of light.

Compare light to a tennis ball. One model that the *Optics* used that comparison. Having entered the prism at right angles to the surface, the beam of light goes through and impinges on the lower or farther side of the prism at a different angle. Imagine that this surface on the farther side is a flimsy cloth penetrated by the tennis ball. The cloth intercepts the ball at an acute angle. The side of the ball that touches the surface before it penetrates will be resisted by friction. The other side, which is not in contact with the cloth, will not be so opposed. Consequently, the ball will acquire a spin.

We can understand this better if we imagine hitting a tennis ball with a racquet. If the racquet strikes the ball in a certain direction, but its face is angled off the path of the strike, the tennis ball will acquire a spin. If we now look at the diagram of the prism, we find that the angle is slightly different at the two ends of the beam of light and must vary continuously between them. Light at one end will acquire a different spin from light at the other end of the beam, with gradually varying levels of spin between the two.

Light is, of course, not a tennis ball but a pressure. The pressure is felt when the solid or rigid space around it touches the eye. It is also like a walking stick touching the hands of the blind man. Just as pressure is felt in the eye as a tendency to motion, so is the spin felt as a tendency to twist. Because there is a different tendency to spin at each point in the beam, the eye feels twists at each point as differently colored images. The variegated colors of the spectrum are states of white light that are felt by the eye as twists when light emerges from the prism. That was his model for prismatic colors.

In the case of a rainbow in the sky, we see one reflection coming off the back of the water droplets when the droplets are at a certain angle. Each droplet at that angle reflects sunlight by a total internal reflection. The angle of its entry and egress determines this. We see all the droplets at any one angle as shaped like a bow: in a bow, droplets showing one color are all at that angle. Of course, all the rainbow colors will appear at their slightly different angles, so we see the whole rainbow with the full array of colors. It is hard to decide which is prettier, the rainbow or the mathematical calculation.

We know the *sine* law of refraction; we have the index of refraction from water to air (and vice versa.) Using this knowledge, we can calculate that there are two and only two angles at which there are total internal reflections of light in a sphere of water. One of these events happens when light enters a droplet of water by refraction, then reflects internally, and then is refracted back into the air to the eye. Since internal reflections require an acute angle, this can be calculated to happen at about 42 degrees.

We can calculate and find that light refracted into a droplet can also be reflected off the back surface of the droplet twice and then refracted out of the droplet to the eye. In every reflection, there is an inversion. When there are two reflections, we will find the spectral colors inverted from the first. This case of double reflection can be calculated to happen at about 52 degrees. We have, therefore, two angles, 42 and 52 degrees, at which total internal reflection takes place. No other angle for a total internal reflection is geometrically possible.

When there are two reflections and two refractions, there is more loss of light than in the case of one reflection and two refractions. Each change in direction will make the image a bit faint, so two reflections make it more faint than one. We can thus understand why the rainbow which appears in all the water droplets at 52 degrees will be faint compared to the rainbow at an angle of 42 degrees, with only a single reflection. All of this appears to be well explained.

The double rainbow is a lovely example of using a hypothetical model to make sense of a phenomenon. Of course, no hypothetical model can be known to be true of reality. We can judge it solely from its effects. Another model may just as well also explain the same effects, for all we know. The success of this model is impressive all the same. Perhaps it is more to the credit of his ingenuity than the method he espoused. But any such success is always counterbalanced by the thought that we do not know the causes of the effects. We have only opined or guessed at them, however impressive they may seem in the writings of an inventive and meticulous natural philosopher.

On the other hand, we have a Cartesian counterclaim: the model does not matter. We know the Principles of Motion as clear and distinct ideas. They are known beyond doubt (he supposed.) To see how such mathematical principles illuminate the phenomena is knowledge enough.

Note

1. Quotes and figures from Descartes' *Dioptrique* are taken from Paul J. Olscamp's translation (Descartes 2001/1637).

References

Descartes, Ren. (1637) 2001. *Discourse on Method, Optics, Geometry, and Metereology*. ed. Paul Olscamp. Indianapolis, Indiana: Hackett.

Newton, Isaac. 1672. New Theory of Light and Colors, Sent by the Author to the Publisher from Cambridge February 6 1671/72. *Philosophical Transactions* 6 (80): 3075–87.

Sabra, A.I. 1967. *Theories of Light from Descartes to Newton*. London: Oldbourne.

12

Applying The Experimental Philosophy

I

Isaac Newton wrote a letter on light to the Secretary of the Royal Society dated February 6, 1672. It provides an excellent Baconian or inductive contrast to the Cartesian method of hypothesis—i.e., to proving first principles first and then shedding light on phenomena with helpful models. Isaac Newton's optics is an example of the "experimental philosophy" adopted by founding Fellows of the Royal Society, such as Robert Boyle or Henry Power. Isaac Newton argues experimentally in his letter, i.e., from beleaguered phenomena as a puzzle to a model of light that gets behind the phenomena. The model extracted is unlike a diagram or a Cartesian aid to the imagination. He describes a formerly hidden reality that he reveals by an "argument"—the argument being a decipherment of a bit of reality behind the appearances.

His technique consisted of picking up an apparent oddity in the appearances and, after collecting more such discrepancies, he resolves all the errors in one latent (hidden, occult, theoretical, or nonevident) schematism or model. When we pay attention to the many oddities in optical

J. Hattiangadi, *Francis Bacon's Skeptical Recipes for New Knowledge*,
https://doi.org/10.1007/978-3-031-52585-8_12

phenomena, we recognize how unshakeable the model is as a solution to the riddle among the appearances.

Isaac Newton wrote to Henry Oldenburg that he was working on grinding a non-spherical lens when he undertook these experiments. As a background to this letter, we should note that a little over a year before the date of this letter, Isaac Newton presented a reflecting telescope to the Royal Society. The Society had received it as a contribution, for which the Society awarded Isaac Newton a coveted Fellowship. Reflecting telescopes avoid an inconvenient limitation of refracting Galilean telescopes. The images of stars in a reflecting telescope do not show colored fringes that blur them, called "chromatic aberration." Therefore, this letter reminds members of the Royal Society about his contribution to the first reflecting telescope ever made and that he was grinding lenses to solve another limitation of telescopes, spherical aberration, in the hope of avoiding another limitation of telescopes. In doing so, he reports a theoretical discovery that led him to build a reflecting telescope. We note here the nod to Baconian *power over nature*.

"I procured me a Triangular glass-Prisme, to try therewith the celebrated *Phenomena of Colours*. And in order thereto having darkened my chamber, and made a small hole in my window shuts, to let in a convenient quantity of the Sun's light, I placed my Prisme at its entrance, that it might thereby be refracted to the opposite wall." He is investigating how a prism produces color, you might think. But he is not. He may get around to it, but mainly he is concerned with geometrical optics. "It was at first a very pleasing divertissement, to view the vivid and intense colours produced thereby; but after a while applying myself to consider them more circumspectly, I became surprised to see them in an *oblong* form; which according to the received laws of Refraction, I expect to have been circular" (Newton 1672, 3076).

Let us pause here. A *circular* beam of light was producing an *oblong* image. The result violates some simple geometrical laws at the angle and distance Isaac Newton had placed them. A circular beam at the appropriate angle of the screen must project a circular image. Of course, the first thought that comes to mind is whether the experimental arrangement is improperly set up. He had studied Isaac Barrow's optical lectures in Cambridge. He knew his geometrical optics, as we might expect of a consummate geometer. He was aware of the importance of arranging all the parts of the apparatus as exactly as possible—as geometry requires. Later in the letter, he even tells us the relevant dimensions exactly, from which an interested reader could calculate whether the image ought to have been circular according to the received theories.

The difference between what the method of hypothesis requires of us and what the experimental philosophy does should be noted. We are witnessing a refutation of the laws of geometrical optics, as Isaac Newton reports. It if we are dealing with an optical theory as a hypothesis, we would ask ourselves, "What is at fault? Is it the law of the conservation of the quantity of motion? Is it the law of the independence of the components of motions to one another?" Or, if we need a new hypothetical model or derivation for the law of refraction, is one of those models at fault? But this is not what Isaac Newton does. He notes the discrepancy in appearance (an apparent application of the law of refraction combined with an evident refutation of it). He collects this *deviant instance*, as Francis Bacon calls it (Bacon, II.29), and then describes the anomalous images obtained by his apparatus.

> They were terminated on the sides with straight lines, but at the ends, the decay of light was so gradual, that it was difficult to judge justly, what was their figure; yet they seemed *semicircular*. (Royal Society. Newton 1672, 3076)

The length of the colored spectrum was about five times the width, and he was "excited" to search for what caused this discrepancy (N. B. not depressed). He then displays his craft of experimental philosophy. He wondered if different glass thicknesses in the prism could explain the difference. So he tried to send the light through different parts of the prism. It did nothing to the elongated image. He also tried to change the hole in the window shutter many times (making it smaller and larger) to see if that made any difference. It did not.

He also wondered if the glass prism he used had some unique defect—*unique* instances are discussed in Bacon, II.28. Isaac Newton must have procured himself another prism for this purpose because he "...took another Prisme like the former, and so placed it, that the light, passing through them both, might be refracted in contrary ways, and so by the latter returned to that course, from which the first had diverted it." (Newton 1672, 3076). The image he obtained was "orbicular"—just as if no prism had been placed in the path of the light. No "contingent irregularity" of the glass could account for the error in the shape of the image. He, therefore, had no *unique* instance to report if we use Baconian terminology again. The discrepancy was genuinely optical and not accidental. Many of us used to the empirical method may suppose that he noted that the fact was dependable and free of bias. But he was doing something else altogether—trying to understand an oddity, an impossibility. We note that his focus is not just on the facts of the matter but on the discrepant facts of the case. He knew there was an Idol or two at work. Which one had to be worked out of the experiments.

Could there be different kinds of sunlight? He examined how the light coming from the different parts of the sun may affect the image differently. Were it true, it would certainly be interesting path to pursue. It would have been what Bacon calls a "revealing instance." (Bacon, II.24).

Before describing his procedure for checking the light from different parts of the Sun, he thoroughly described all the critical features of the experiment. He carefully "measured the several lines and angles, belonging to the Image." Any Baconian experiment must have the relevant conditions to replicate the experiment. Isaac Newton provides them here. His report is detailed. The details are not pertinent to our purpose, which is to see his method illustrated, but some are instructive.

> The distance of the Image on the wall from the hole or the Prisme was 22 foot; its utmost length 13¼ inches; its breadth 2⅝ inches; the diameter of the whole ¼ of an inch; the angle, which the Rays, tending towards the middle of the image, made with those lines, in which they would have proceeded without refraction, 44 deg. 56′. And the vertical angle of the prism, 63 deg. 12′. (Newton 1672, 3077)

He writes that the angles on either side of the prism were "as much as I could make them equal, and consequently about 54 deg. 4′." Equality of their angles is a vital notation because, other than in this specific position, the image will not be circular according to the laws of optics. The rays of light must fall perpendicularly to the wall. Calculating the angles of the image as subtended from the hole, he shows that about 31′ angle is answerable to the sun's diameter, which was equal to the width of the image, but the length subtended was more than 2 deg. 49′. As geometrical theory predicts, the angle of 31′ (its width) should also subtend the same length. But this length was so much greater.

We note that by a simple calculation, he has shown that the width is not but the length is anomalous. An unaccounted angle remains at 2 deg, 18′ along the length. Again, we note that he was not accurately measuring a perception, a law, a regularity, or a reality. Precision was not his aim. He was measuring the oddity of a funny shape, and by calculation, he located the irregularity in the length but not in the width of the discrepant image.

> Having made these observations, I first computed from them the refractive power of that glass, and found it measured by the *ratio* of their sines 20 to 31. And then by that *ratio*, I computed the refraction of two Rays flowing from opposite parts of the Sun's *discus*, so as to differ 31′ in their obliquity of Incidence, and found, that the emergent Rays should have comprehended an angle about 31′, as they did, before they were Incident. (Newton 1672 1672, 3077)

He was using the Cartesian *sine* law of refraction here, and it is worth noting what he says about it: "But because this computation was founded on the Hypothesis of the proportionality… [of the *sine* law] … which… I could not imagine to be so erroneous as to make that Angle but 31′ which in reality was 2 deg. 49′; yet my curiosity made me again take to my Prisme." (Newton 1672, 3077).

For the time being, he had accepted the result of the *sine* law as a "hypothesis" that many would have considered a demonstrated mathematical law. Here we find the key to the experimental philosophy: take every apparent law hostage, treat it as a mere hypothesis, but do not give it up as entirely false. Unlike Francis Bacon's initial suggestion to avoid all hypotheses, the experimental philosophy allowed experimenters to entertain hypothetical principles. Isaac Newton himself believed, he admitted, that the difference of 31′ deg and 2 deg. 49′ in angle cannot be due to a defect in the hypothesis of the *sines* of refraction, which law he had himself tested. He "takes to his prism" to see if he could detect if the rays from different parts of the Sun could be so different as to get them to diverge in length of the image. He was building up an experimental and natural history.

He rotates the prism slightly about its long axis, up and down (rotating it along the direction of the length of the image on the wall), by an angle of about 4 or 5 degrees, trying to find out if this makes the image change much in length. By rotating it thus he could catch light from different parts of the sun on his prism. But he found that "the Colours were not thereby sensibly translated much from their place on the wall," which showed him that changing the Incident light from the upper part of the sun or the lower did not make a difference in the relation of the place of the colors in the image on his screen. He reports that he had nothing to report. However futile in this instance, the technique of rotating the prims comes in useful later in his hunt.

"Then I began to suspect whether the Rays, in their rejection through the Prisme move in curved lines…." If the light were to travel in varying curves, as it traversed towards the image, then the curvature could explain the discrepancy length, of course. Then there would still be the problem of why the light curves. Here he invokes the Cartesian hypothesis of light as a tennis ball that has a curve in its path. The curve on a ball could make it swerve due to spin and the consequent differential friction in the air. Isaac Newton writes, "when I came to examine it, I found no such curvity in them." (Newton 1672, 3078).

"I observed, that the difference betwixt the length of the Image and the diameter of the hole, through which the light was transmitted, was proportionable to their distance." Moving a screen back and forth in the direction between the image and the prism, he saw at different distances that the length of the image changed proportionately to the distance from the source. If there were curves, the length of the image would increase disproportionately as the distance from the prism increased or decreased. It did not.

"The gradual removal of these suspicions at length led me to the *Experimentum Crucis*, which was this:" He does not explain why he called this experiment crucial. In Francis Bacon's list of privileged instances, he mentions an instance of the cross or of the fingerpost (Bacon, II.36). This instance is like a sign showing us how to proceed when one reaches a crossroads. Isaac Newton does not describe the crossroads he has reached, but we can guess that the choice is this: Is the experimental arrangement or another contingent fact responsible for the elongated image? Or is it the law of refraction itself creating the problem here? He notes that he has tested the law outside of this experiment, and found it works tolerably well.

The famous crucial experiment is this: he took two boards with a small hole in each and placed them 12 feet apart, with one board behind the prism near the window to let light pass through it. The image from the lens should fall upon the other board 12 feet away, covering the hole made for the purpose. Some of the light from the image should then also pass through the second hole. Now he placed a second prism beyond the second hole to receive the small amount of light passing through it as an incident ray to examine its refraction on the wall. He took hold of the first prism beside the window and rotated it slightly around its axis (as before.) Thus, he passed the isolated light in various parts of the image through the second prism. In effect, he was letting different colors selectively through the second hole.

Isaac Newton did all this "that I might observe," he wrote, "to what places on the wall the second Prisme would refract them." He found that the light from one end of the image on the second board was refracted at a different place than the light from the other end of the prism. "And so the true cause of the of the length of that Image was detected to be no other, than then light consists of Rays Differently Refrangible, which, without any respect to a difference in their angle of incidence, were, according to their degree of refrangibility, transmitted to different parts of the wall." (Newton 1672, 3079)

His last experiment above does, indeed, show that different parts of an image will refract through a second prism at different angles, even though they have the same angle of incidence at the second prism. They are "differently refrangible." It follows that sunlight is not homogenous but consists of mixed rays, each with its refrangibility—and a correlated color!

Heterogeneous light will, therefore, limit the precision of an image of any star when seen through a refracting telescope. The image will be blurred to the extent of the differences in the refrangibility of different rays within white light, which is why Isaac Newton built his famous reflecting telescope. Reflections are uniform in this respect: the angle of incidence and reflection are always equal. Notice how he got a bit more "power" over the image in a telescope to express it in Baconian terms. This power over nature was not as useful as he had hoped because the reflecting telescope did not find much subsequent use. Improvements in refracting telescopes were also soon made. The main advantage conferred by the reflecting telescope was not in its use but in its "pledge of truth."

Historians and philosophers are quite convinced nowadays that the inductive method does not work. Isaac Newton, we suppose, could not "deduce" or "inductively infer" (his occasional words) the mixed rays of sunlight from a mere experiment. Whatever may be that "experimental philosophy," which the Royal Society proclaimed, we think that it is not and cannot be a viable method. We *must* assume a hypothesis somewhere, we suppose with our current methodological sophistication, to get a result of any generality. But is a mistake to focus on the use of hypotheses. For the experimental philosophers produce models, when successful, that are unconditional—not at all hypothetical. In the process, hypotheses may be used. But their role is to disappear.

Examine Isaac Newton's report of his investigation of the anomalous shape of the image. He conducts a series of tests of varying hypotheses. Aha! (We say). This philosopher is sneaking them in! But if we notice more closely, the hypotheses are serially discarded, or refuted, but something else is happening in the background. The phenomena adduced are building up into a knotty, enigmatic natural history. At the end, when he announces that white light has all the colors of the spectrum hidden in them, we cannot resist the conclusion. It is this fact, that the conclusion becomes unshakeable, which marks it out as a successful experimental result. Loosely, he can claim to have "deduced" it from the phenomena themselves.

The experiment above shows us that this is a mistaken assumption. Isaac Newton focuses fiercely on resolving a discrepancy by finding why it happens. He uses what we may call an error analysis. He shows little interest in testing general hypotheses except to generate discrepancies in appearance—leading to more experiments. Ultimately, the discrepancy must disappear, and the appearances must be made coherent.

Even plainly written texts become difficult to interpret if we assume that what we see him doing before our eyes is impossible. It seems to us today that induction is impossible, as Pierre Duhem argued, though he did this when discussing the theory of universal gravitation. We will not take gravity up in this volume but leave it for a sequel. Not allowing for its very possibility, following Pierre Duhem, we are led to assume that there must be some sleight of hand, a hidden assumption. Our sophistication concerning method has led even so careful and thorough a scholar of Isaac Newton's optical theories as Alan Shapiro to question Isaac Newton's inductive thought in optics.

Alan Shapiro (1990) has raised an interesting question: Why did Isaac Newton not use the phrase "experimental philosophy" in his published writings until early in the eighteenth century? And when using those words, why did he use them only as late as 1713, in the second edition of his *Principia*, to defend his theory of gravity as not being a hypothesis? Perhaps the phrase "experimental philosophy" was not a plain description of his method. Invoking that phrase later could have been a rhetorical ploy to which he resorted. Perhaps it was a change in methodology. Perhaps it was merely designed to avoid criticism in his celebrated dispute with Gottfried Leibniz (Alexander 1956). All these possibilities arise. Were Alan Shapiro right, we would need to revise our opinion of the meaning of the plain text we just read. In the 1672 letter, it is true, he does not use those words. In his letter to Henry Oldenburg of February 6, 1672, Isaac Newton said this after the above experimental solution was reported. Alan Shapiro also notes it, as we would fully expect from a careful and renowned scholar of Isaac Newton's optics.

"I shall now proceed to acquaint you with another more notable deformity in its Rays, where the *Origin of Colours* is unfolded. A naturalist would scarce expect ye science of those become mathematical, & yet I dare affirm that there is as much certainty in it as in any other part of Optics. For what I shall tell concerning them is not an Hypothesis, but most rigid consequence, not conjectured by barely inferring 'tis thus because not otherwise or because it satisfies all phenomena (the Philosophers universal Topick,) but evinced by ye mediation of experiments directly and without any suspicion of doubt."[1] We note that he is all but calling it a Baconian or experimental induction.

We are inclined by two hundred years of philosophical argument to assume that he must be incorrect in his claim to have made a direct derivation from experiments as he claims. Let us assume that he is mistaken. For the sake of this argument, we can doubt whether his inductive claim is valid. Even if he were mistaken that he derived his theory of the heterogeneity of white light "by ye experiments directly," it does not affect his intention to do so, nor his meaning. He says he was doing experimental philosophy as he understood it. He may have done it badly, if that is what we judge. It does not follow that he was not trying to derive the theory of white light from his experiments. He does say so. Consider a similar case.

If the Cartesian *Cogito* argument fails somewhere, we would not conclude that René Descartes was not seeking an indubitable foundation for philosophy. It fails in many ways, according to later philosophers. But we do not deny what he was trying. For his part, Isaac Newton's thirteen propositions that followed the experimental argument above were also part of what he understood as "experimental philosophy." It is hard to understand what he is saying if he is not signaling that he is practicing that very experimental philosophy that the Royal Society was formed to encourage.[2] Moreover, Isaac Newton had just become a Fellow of that august body the year before after presenting his reflecting telescope. He was a novice. Did he need to show the Fellows of standing how he deduced the theory behind the facts of the experimental discrepancies?

Isaac Newton describes the experimental philosophy forty-one years later, in a letter of 1713 to Roger Cotes, the editor of the second edition of his *Principia*. Isaac Newton wrote, "whatever is not deduced from the phenomena must be called a hypothesis; and hypotheses, whether metaphysical or physical, or based on occult qualities, or mechanical, have no place in experimental philosophy. In this philosophy, propositions are deduced from the phenomena and are made general by induction."[3] This quotation matches what he said in 1672.

Alan Shapiro is a careful scholar, and he is correct that the phrase "experimental philosophy" does not occur in the 1672 paper under discussion. However, all the facts adduced by Alan Shapiro can be subsumed under another explanation than his. For an excellent reason, Isaac Newton did not need to use the phrase when talking about it to the Fellows of the Royal Society in 1672. He was then a neophyte at experimental philosophy and not a Fellow of the distinguished Society for long; he was showing them what he had learned from them in their publications, that is to say, *from their practice of experimental philosophy*. Did he need to tell them what it is or show them that his optics is an example? For instance, he did not assert that his experiments were composed in English prose. He expected everyone to know it. He did not write that he depended on Euclid's geometry to understand straight lines or Apollonius to apprehend the curvature of conic sections. Everyone could grasp such facts if they knew their geometry. He did not need to tell them that he was doing experimental philosophy. He could assume they knew it. Of all the people in the world, the Fellows of the Royal Society knew it. We do not know it today because, as methodologists and philosophers, we know so much more sophisticated methodology that we have forgotten what it is.

Much later, in 1713, Isaac Newton realized he had to explain his actions when he was engaged in a dispute concerning gravitation with Gottfried Leibniz and others. While many, Gottfried Leibniz included, had allowed the heterogeneity of sunlight to be extracted from experiments, they did not allow the same status to universal gravitation. That was going too far. Most Cartesians questioned it then, and it remains a common doubt among philosophers of science today. Isaac Newton needed to remind everyone in 1713 that his philosophical method was still experimental and not Cartesian. By contrast, Gottfried Leibniz was raising criticisms as a fellow rational mechanist. Isaac Newton was also a reformed Cartesian in many respects. But he rejected the demand for a Cartesian explanation of forces. There was no adequate mechanical explanation known to him. That was not the point of his *Principia*.

There is a limitation of the method of experimental philosophy. We cannot go beyond the model discovered. We found whatever we have found with great difficulty, extracting the schematism from the distortions and discrepancies in the historical record of facts. Discovering an even deeper principle to explain the new model will need a new experimental history to compel it. If we are ever asked to explain the just revealed model, we must insist that nothing as good is yet available. The revelation contained in a riddle's solution is our best knowledge. To add explanations of the underlying nature of the things (e.g., forces) is to engage in what Francis Bacon called "Anticipating Nature." Anticipations, or even grand "first" principles, had been admitted by Robert Boyle into experimental philosophy but only as hypotheses. The corpuscular philosophy was what we might call a "working" hypothesis, which is disposable when the puzzle demands it. It is not an experimental result. But they do not appear in the final model "deduced from the phenomena." Later, Isaac Newton would say that we do not feign hypotheses in experimental philosophy. He should have added, "in or final models."

The Baconian method may be compared usefully to a limitation of the Cartesian method. In the Cartesian method, the principles of philosophy are well-founded. They are known for sure. René Descartes may have made a mistake in his arguments and enunciated false principles. That is merely a personal flaw. The principles of philosophy were conceived as known with absolute certainty. But, no matter how indubitable, their application to particulars—that is to say, shedding light on them by employing a model of the phenomena—remains hypothetical and uncertain. Being hidden from us, the reality proposed in a model is almost always invisible. Innumerable hypothetical models may be conceived, and we may say the one we choose to endorse is "conjectured by barely inferring 'tis thus because not otherwise or because it satisfies all phenomena (the Philosophers universal Topick)" in Isaac Newton's laconic words.

Isaac Newton gives us a fair description of the hypothetico-deductive method we currently admire. He sees that generating Cartesian knowledge of any hidden reality is futile. Christiaan Huygens, that great contemporary Cartesian whom he admired, had proposed that the best hypothetical model may be ascertained by using a mathematical calculation of probability from the collection of sensory phenomena. The mathematical theory of probability was based on finite games of chance. It was further developed as a mathematical theory by Blaise Pascal and Pierre de Fermat. Modern philosophy of science was just starting, with all the controversies and doubts about method to follow. Isaac Newton dismisses it in a few brief sentences, because, like Robert Boyle, Robert Hooke and other Fellows of the Royal Society, he was following what seemed clear to him (though not to us)—the experimental philosophy.

A hidden reality in experimental philosophy is inferred directly from the problematic phenomena. The model that is faithful to reality is a uniquely coherent model that makes plain (or explains, makes coherent) many logically impossible problems contained in the puzzle. The solution presumably gives us knowledge since the solution is unique in bringing coherence to our puzzling beliefs. But there is no immediate path from models to first principles. We need new experimental problems yet to be recorded to go up the pyramid of science, as Francis Bacon envisaged it.

The matter is worse still because the experimental method is fallible. Our models may be upended because Isaac Newton made a personal error but also because, though he made no error, the puzzle grows in size after being solved. An old solution may not remain a coherent solution to an expanded puzzle. An expanded puzzle supersedes the old. A new solution may be found to it that improves on an old solution to an old puzzle. The heterogeneous composition of white light has remained with us as a piece of optical knowledge, but not Isaac Newton's theory of gravity. The latter was also an experimental derivation, as he claimed. But it has been replaced by a better model early in the last century. With general relativity, Albert Einstein solved a larger puzzle with even more problems.

II

The main claim to solving the riddle of colors did not come merely from his constructing a model of light from the *experimentum crucis*. The model also had to make all the other problems of color go away. There were many of those recorded by Robert Hooke and Robert Boyle. The new model solved the puzzle of the experimental and natural history of colors. It was carefully assembled.

Isaac Newton went on to discuss colors in his paper of 1672, which he had largely avoided in the first part of his geometrical induction. It is the colors that were paradoxical. The puzzle that he solves is about color. Almost miraculously, by his breakthrough, he brings a new understanding of them. The discrepancy he took up was mathematical. He used a technique important to his experimental argument. Once he saw that sunlight has many different rays mixed within it, he could use that knowledge to find more knowledge, again using experiments:

1. "As the Rays of light differ in degrees of Refrangibility, so they also differ in their disposition to exhibit this or that colour." (Newton 1672, 3081)
2. "To the Same degree of Refrangibilty belongs the same colour and to the same colour always belongs the same degree of Refrangibility…."

Please note that the experiment that shows the correlation of colors to the refrangibility of rays is by variations of the same *experimentum crucis*. When the light of different colors goes through the second prism, they go through different angles. Moreover, each color refracts at its characteristic angle. Just as the *sine* law can be used to measure and find the refractive index of light going from glass into air, so too can we map out the refrangibility of each color. Earlier in the text, he says his conclusions are as sure as any in Optics. We see why.

> The species of colour, and degrees of Refangibility proper to any particular sort of Rays, is not mutable by Refraction, nor by Reflection from natural bodies, nor by any other cause….

The claim made by Isaac Newton above that colors, when pure, remain immutable is a strong claim. At first, he argues that he could not change it no matter how hard he tried. But this is not the only reason for it. His point is that in their mixture in white light, the colors do not blend but remain separated. His experimental result is still important for the theory of light.

Colors can be mixed to produce other colors. Green, for instance, can be produced by mixing blue and yellow. But when this mixed green light is sent through a prism, it resolves back into its original component colors. There are pure colors, however, that retain their color through any refractions or reflections. Green, for instance, is a pure color in the spectrum, and that pure green light remains green when separate no matter what you do to it. Isaac Newton names the pure colors of the spectrum as seven in number: red, orange, yellow, green, blue, indigo, and violet. But there are indefinitely many gradations between them that are also pure colors.

Mixed colors from light near one another produce intermediate colors, like orange from yellow and red, or green from blue and yellow, but when they are far apart, colors do not produce intermediate colors. They produce colors such as brown that are not intermediate. The next two items on his list of thirteen are interesting.

"7. But the most surprising and wonderful composition was that of *Whiteness*. There is no one sort of Rays that alone can exhibit this. 'Tis ever compounded, and to its composition are requisite all the aforementioned primary Colours, mixed in due proportion. I have often with Admiration beheld, that all the Colours of the Prisme being made to converge, and to be again mixed as they were in the light before it was Incident upon the Prisme, reproduced light, entirely and perfectly white, and not at all sensibly differing from the *direct* light of the Sun," he says, provided, of course, that the lens of the prism that we use is clear. He describes an experiment at the end of his list of thirteen theses that exhibits this to be the case (Newton 1672, 3083).

> 8. Hence, therefore, it comes to pass that *Whiteness* is the usual colour of *Light*; For, Light is a confused aggregate of Rays indued with all sorts of Colors, as they are promiscuously darted from the various parts of luminous bodies. And of such a confused aggregate as I said, is generated Whiteness, if there be a due proportion of the Ingredients; But if any predominate, the Light must incline to that color; as it happens in the Blue flame of Brimstone; the Yellow flame of a candle; and the various colours of the fixed stars.

In the eighth part, we already see him applying his knowledge to cases where there seemed to be riddles, and he continues this in the next four propositions. The prism, he says, merely separates the colors by their unequal refrangibility, and similarly in the case of rainbows. Examining the Cartesian explanation, we find that Newtonian induction differs only in that colors are not a state of white light but original to light and mixed within it.

Isaac Newton calls the following results a part of "general induction." In his writing, general induction sometimes sounds like an inessential later step. But we can see that it is central to what Francis Bacon called "induction," in which the interpretation of nature is satisfactorily concluded. All the discrepancies in the "experimental and natural history" have been resolved at this step. Of thin plates like leaf gold, fragments of colored glass, and *Lignum Nephriticum*, which had been described by Robert Boyle, which exhibit one color on one side and another color on the other, he has this to say: they "are on these grounds no longer riddles," he says as they are explained easily by showing how at certain angles one color is reflected and at another a different one, since the refracted light has different refrangibilities mixed within it (Newton 1672, 3084). Once we know that the color of a body is its prediliction to reflect one kind of light rather than the other refrangibilities (which it absorbs) the strange dual colors of thin gold plates of gold and the peacock's feather are not hard to understand. (Thin plates were to occupy his attention later in life, but these were studied by further experiments).

Robert Hooke had recorded another riddle, which was also resolved. A transparent glass filled with only red liquid and another filled with only blue liquid are each transparent enough. Each imbues light passing through it with its hue, but together they are opaque. This is puzzling because any red hue that is made a bluish should turn purple. Why does a bluish red become opaque? It is easily explained on the new model. One glass transmits only red light (the rest being absorbed or reflected) and the other only blue (the rest absorbed or reflected.) But they transmit none when the flasks are placed together in tandem or series. Another piece of the riddle of colors is solved.

The point cannot be overemphasized. Solving the whole riddle makes the model unique. An experimental and natural history records many conjointly impossible facts. We can devise many hypotheses to deal with one or other of these discrepancies. But in the light of other discrepancies, a hypothesis remains unsatisfactory. But by making all the impossibilities coherent in a model Isaac Newton makes the model uniquely coherent. If the coherence of a natural history model is unique—if every other attempt runs into one of the difficulties recorded—then we have knowledge. What he calls "general induction" is essential to his method.

The thirteenth proposition is interesting, for he claims that "the Colours of all natural Bodies have no other Origin than this, that they are variously qualified to reflect one sort of light in greater quantity than another. And this I have experimented in a dark Room, by illuminating those bodies with uncompounded light of divers colours. For by that means any body may be made to appear of any colour. They have there no appropriate colour, but ever appear of the colour of the light cast upon them, but yet with this difference, that they are more brisk and vivid in the light of their own day-light colour." (Newton 1672, 3084). We see that bodies do not possess color, but light does. Bodies exhibit reflectancies.

We have already noted that Isaac Newton first builds a model of light and then (by "general induction") he explains all the known puzzles of the colors. General induction is the Interpretation of Nature, as Francis Bacon had supposed. The more extensive the puzzle, the less doubt we have that the coherence achieved by the general induction is unique. It is stable and constitutes knowledge.

The way Isaac Newton goes about building the model is the first step. He works with the phenomena, experimenting and thinking about them, and this step was an invention of his own. In the case of colors, the steps were small and the results conveyed in a short letter. When he built a model the System of the World, enlisting a new theory of gravity, the first step was much more elaborate. The model building took up several hundred pages of dense mathematical argument, including the Book I of his *Mathematical Principles of Natural Philosophy* and about a third of Book III on the *System of the World.* Only after the model-construction was completed could he use it to solve the riddle of the experimental history created by the Copernican debate. The solution was "by a general induction."

Isaac Newton's model-building technique was innovative. It had a significant influence on later science, and on how the Baconian method, or the experimental philosophy, was applied. But that account is too long to take up here. It goes beyond the scope of this volume to do more than mention it as a topic worthy of study.

Another experiment Isaac Newton reported is very interesting because it taught him something else that is important about white light. As before, send sunlight through a 1/3 inch round hole in a shutter, and pass it through a prism mounted there. Then mount a large convex lens about three feet away in the way of the light with a thickness of three or four inches, making sure that the lens captures all the light. The lens will gather the image and send it to the point that constitutes its optical focus. If we intercept the light with a piece of paper, we find that the light will show the spectral colors before and after the focal point. If experimentally we locate the point or the focus of the lens, the image on the paper will be found there to be perfectly white, just like the sunlight admitted into the room. However, something interesting happens if we intercept one color in the light beam before its incidence into the second lens. If we interpose a piece of paper to block one color, the image at the focus will change from white into the spectrum—but with the intercepted color missing. Then if we remove the interception, the color of the image will become white again.

The sequence of experiments shows us that the colors mixed in white light remain distinct and different. It also shows that the colors we separate into the spectrum can recombine to form the same white light again. To test its power over nature, we should be able to manufacture white light by mixing the right colors in the right proportion. Isaac Newton devised a wheel of proportionally colored lights. When such a wheel is rotated, the colors disappear. Pure white light replaces them.

Once all the known problems are resolved, the Baconain induction is complete. But our curiosity is great. We want to know more. We wish to go beyond the puzzle, the induction. And Isaac Newton was no exception. He also added some hypotheses. His foray into Anticpations of Nature is understandable. All of us do it. Francis Bacon did write that our minds naturally rise, and need leaden weights to hold them down. From the perspective of illustrating his methodology, that is to say, for our purposes, Isaac Newton muddied the waters when he went beyond the model he had discovered. But there are two kinds of hypotheses to distinguish. There are those he introduced because he was curious about light. They were not submitted voluntarily but out of the need to solve a puzzle. They are refuted and completely disappear in the model, i.e., in the conclusion. When he speculated out of curiosity, at the end, he was deceived by some Idols, as Francis Bacon might say. The hypotheses he began entertaining remained hypotheses even after the inductive study was complete.

III

Weavers of carpets and rugs in the past would spend years on a single mat with intricate designs that were consummately woven together without noticeable blemish or error. It is a wonder how they kept their overall design and the constancy of colors in mind as they worked meticulously on detail on a great rug, sometimes for years. But just as they were about to finish, after all their effort at implementing an accurate design, some weavers would leave one blemish in the design in some inconspicuous corner of the rug. It was like a prayer. That deliberate fault signified their simple, monotheistic, devout, faith that God alone is perfect.

The craft of experimental philosophy is different. One does not need to find a place to make a mistake. The errors find you in philosophy. Philosophers have pored over Isaac Newton's experiments. They claimed to have found errors in his conclusions. Some of these errors claimed were imaginary, arising out of misunderstanding. Other errors are genuine. But when Isaac Newton committed errors and was caught out, it was because he was doing experimental philosophy. That is how the craft works. You learn from a compilation of errors by an analysis. In experimental philosophy, to miss a mistake or deception in such an analysis is not a surprise but an expectation. Philosophy is fallible. Errors are many, and solutions are contingent on the few you have found.

Christiaan Huygens, the contemporary philosopher perhaps most admired by Isaac Newton, found an error in the paper. Isaac Newton had assumed in his letter on light and color that all white light is like sunlight and would exhibit all the colors of sunlight in appropriate proportions. His wheel showed how combining all the colors in the right proportions would generate white light when the wheel is spun. He had demonstrated all the colors of which white light is composed. However obvious this may seem, it turns out that it is not the case. Three primary colors (red, blue, and green) are enough to produce white light if mixed in due proportion. When he published his *Optics* in 1704, Isaac Newton had to concede this.

Now we are apt to say that light that appears white to us depends in part on our physiology, on the rods and cones on our retina. Isaac Newton avoided this topic as a hypothetical. "But, to determine more absolutely what Light is, after what manner refracted, and by what modes or actions it produced in our minds the Phantasms of Colours is not so easier. And I shall not mingle conjectures with certainties." (Newton 1672, 3085). So he avoided hypotheses. But experimental philosophy, unlike the certainty promised by Cartesian first principles, only ever promises relatively uncertain or fallible knowledge. The blemishes in your rug arise despite your best efforts. They seek you out.

Robert Hooke was another notable colleague, a member of the Royal Society, with whom Isaac Newton exchanged letters on this topic. There was a cordial exchange between them some years before their famous quarrel. Robert Hooke proposed a theory of light and color based on light as a vibration. According to his (modified Cartesian) theory, the light had two colors within it and acquired the various colors by the action of the prism affecting the state of the light passing through it. He found that he could explain most of the phenomena that Isaac Newton had described, but could not deal with them all.

The exchange between them is interesting because we see more clearly how carefully Isaac Newton avoided hypotheses to stay within the knowledge that experiment allows. Isaac Newton noted that his conclusion that white sunlight contains a mixture of all the colors would be compatible with a vibrational hypothesis of light if those vibrations did not blend in with one another in white light. For if they blended and then separated again into the same components, there would need to be some mechanism for it to happen so exactly. After all, experiments had shown that colors remained unblended in white light.

As opposed to the vibrational model, Isaac Newton had proposed another hypothesis, an additional hypothesis about light. Let us quote Isaac Newton first on his reading of his experiments and why he thinks it is not just any hypothesis. "…[I]t can be no longer disputed…, perhaps, whether Light be a Body." Note the word "perhaps"—it shows that he is not asserting it, just considering it. "For when Colours are the *qualities* of Light," [we see that since they are not in bodies, which only reflect colors, but do not possess them] "having the Rays for their Entire and immediate subject, how can we think those Rays *qualities* also, unless one quality may be the subject of and sustain another; which in effect is to call it *substance....*" (Newton 1672, 3085)

Isaac Newton has a point, of course. He has shown that whatever colored light is, it exists in white light, and the prism merely separates preexisting colors. It does not seem to him at all reasonable that light could be anything but a substance or a thing (a particle?) How else could a ray of one color retain its independence from another if both are a kind of motion and not separate bodies? He is entertaining some such Aristotelian thought.

But this is not experimental philosophy at work. Isaac Newton is relying on a principle about the difference between a substance and a quality, which is not extracted from his experiments, but introduced *ex cathedra*. By his standards, it is a hypothesis. After all, even the *sine* law is a hypothesis. Moreover, corpuscularism is not a hypothesis that his experimental philosophy presupposes. It is added on, or assumed. We can take it out of his letter on light and color, and the rest of it stands unaffected. Later in his life, a more mature Isaac Newton would have asked this sort of thing as a *Query*. He would not assert it as a claim. But it shows he is not kindly inclined to a wave theory of light.

What he has to say about Robert Hooke's vibrational hypothesis is quite interesting, though."....Mr. Hooke's *Hypothesis* as to the fundamental part of it, is not against me. The fundamental supposition is, that the parts of bodies when briskly agitated, do excite vibrations in the aether, who are propagated every way from those bodies in straight lines, and cause a sensation of light by beating and dashing against ye bottom of the eye, something after the manner that Vibrations in the air cause a sensation of Sound by beating against the Organs of hearing. Now the most free & natural application of this Hypothesis to the solution of Phenomena," [please note that phrase, "solution"] "I take to be this: That the agitated parts of bodies according to their several sizes, figures & motions, excite vibrations in the Aether of various depths and bignesses," [we note that this passage is a description of wavelengths,] "which being promiscuously propagated throughout the medium to our eyes, effect in us a sensation of light of a white colour; but if by any means those of unequal bignesses be separated from one another, the largest being a sensation of Red colour, ye least or shortest of deep Violet, the intermediate ones of intermediate colours…" Note how he gives a pioneering and sympathetic wave account of the phenomena of mixed colors on behalf of Robert Hooke's hypothesis. Then he goes on to show how, if the waves stay separated in white light, they can explain all the phenomena he has adduced so far (11 June 1672 letter to Henry Oldenburg, in Newton 1959, 174–75)

"And they agree so justly with my Theories that if Mr. Hook think it to apply them, he need not on that account feared a divorce from it." So his point is that what he finds in his experimental philosophy must be taken into account by any hypothesis like Robert Hooke's. Isaac Newton's reason for rejecting a wave theory of light is that it cannot explain how light can travel in a straight line if it is a wave. "For to me the fundamental supposition itself seems impossible; namely that the waves or vibrations of any fluid can like the rays of Light be propagated in straight lines, without a continual and very extravagant spreading and bending every way into ye quiescent Medium where they are terminated by it." (Newton 1959, 175)

Isaac Newton makes two mistakes here. He is arguing for "'tis thus because 'tis not otherwise, or because it satisfies all the phenomena (the Philosophers universal Topick)" doing exactly as he said he would not. It must be a body because it cannot be a wave and it satisfies all the phenomena.

Well, he was not perfect.

His second mistake was to suppose impossible what merely appears to be impossible, namely that waves of light can appear to travel in a straight line. An experimental philosopher is a skeptic and should not assent so easily to what such arguments claim. In the vibrational theory of light, rectilinear propagation can be explained as an *appearance*. The mathematical hypothesis of transverse waves proposed by Augustin Fresnel in the nineteenth century implied something unexpected. Light does not travel in rectilinear fashion, but it does appear rectilinear on a first impression. Primary waves must be distinguished from secondary waves, as Christiaan Huygens argued. A more careful study of the phenomena shows that secondary light waves bend into shadows. Shadows are places where the secondary waves cancel each other out, as only waves can do. Bodies or substances do not enhance or obliterate other bodies, but waves do it all the time, depending on whether waves interact constructively or destructively in the shadow.

Place a spherical object in the path of a parallel beam of light. You will see a shadow with a light spot at its geometric center. A shadow is where the (secondary) light waves cancel out, not where the light does not reach. Super positioning of waves to produce shadow was Christiaan Huygens' insight. The light spot was mathematically derived from Augustin Fresnel's equation by Siméon-Denis Poisson, who thought it an absurd consequence. François Arago then found the "absurd" spot experimentally. We could name that bright spot after any of the four great natural philosophers involved. It is usually named Arago's spot among experimenters and Poisson's among theorists.

New errors find us out. We say we find them. Usually, it is a critic who does. The appearances betray our best insights. Even after a consummate error analysis of known errors, which is a fair description of Isaac Newton's discovery of heterogeneous white light, there were more errors to discover. Not all were his to find. Unlike the weaving monotheist, you do not need to weave a deliberate fault in an inconspicuous corner of your rug. In skeptical philosophy, your errors catch up with you.

Notes

1. From Isaac Newton (1672, 3081), with an emendation that can be found in the version found in Isaac Newton (1959, 97).
2. Kirsten Walsh in (2021) and other papers argues that from Isaac Newton's writing we cannot discern a change in methodology beginning with his optics in 1672 and going all the way to his physical astronomy as revised up to 1726. His oft quoted remark "hypothesis non fingo" is an important part of his laconic remarks on method. However it was made, in the second edition of his *Principia,* in 1713, which is much later than his optics and often supposed to be a different method. His method was poorly self-described, and Alan Shapiro (2021) is right that his later gloss on it betrays an empiricist concession missing in his early thoughts on method. But Kirsten Wlash is right at a more basic level—his craft and his use of the experimental method did not change through his long career. His method

in his practice will be discussed in a sequel to this book, currently entitled *Isaac Newton's Experimental Craft Hidden by the Pax Empirica.*
3. From Isaac Newton's letter dated 28 March, 1713 to Roger Cotes (Newton 1975, 397).

References

Alexander, H.G., ed. 1956. *The Leibniz-Clarke Correspondence.* Edited by H.G. Alexander. Manchester: Manchester University Press.

Newton, Isaac. 1672. New Theory of Light and Colors, Sent by the Author to the Publisher from Cambridge February 6 1671/72. *Philosophical Transactions* 6 (80): 3075–3087.

———. 1959. *Correspondence of Isaac Newton, Volume 1 1661–1675.* Edited and Translated by Herbert W. Turnbull. Cambridge: Cambridge University Press.

———. 1975. *The Correspondence of Isaac Newton, Volume 5 1709–13.* Edited and Translated by A. Ruper Hall and Laura Tilling. Cambridge: Cambridge University Press.

Shapiro, Alan E. 1990. *Before Newton: The Life and Times of Isaac Barrow.* Cambridge: Cambridge University Press

Shapiro, Alan. 2021. Newton Did Change His Views from Certainty to Probability. *Philosophy of Science* 88 (1): 169–180.

Walsh, Kirsten. 2021. Newton's Scaffolding: The Instrumental Roles of His Optical Hypotheses. In *Experiment, Speculation and Religion in Early Modern Philosophy*, ed. Alberto Vanzo, Peter Anstey. New York, NY: Routledge.

References

Alexander, H.G., ed. 1956. *The Leibniz-Clarke Correspondence*. Manchester: Manchester University Press.

Anstey, Peter. 2014. Philosophy of Experiment in Early Modern England: The Case of Bacon, Boyle and Hooke. *Early Science and Medicine* 19 (2): 103–32.

Aristotle. 1984/4th c. BCE. *The Complete Works of Aristotle: The Revised Oxford Translation*, 2 vols. Ed. Jonathan Barnes. Princeton, NJ: Princeton University Press.

Aubrey, John. 1898. *Brief Lives, Chiefly of Contemporaries, Set Down by John Aubrey, between the Years 1669 & 1696*, 2 vols. Ed. Andrew Clark. Oxford: Clarendon Press.

———. 2015. *Brief Lives with an Apparatus for the Lives of Our English Mathematical Writers*. Ed. Kate Bennett. Oxford: Oxford University Press.

Augustine, Bishop of Hippo. 2019/386. *Against the Academics: St. Augustine's Cassiciacum Dialogues, Volume 1*, New Haven, CT: Yale University Press. https://doi.org.ezproxy.library.yorku.ca/10.12987/9780300244878.

Bacon, Francis. 1626. *Sylva Sylvarum, or A Naturall History in Ten Centuries*. Edited by William Rawley. London: William Lee.

———. (1603) 1857. Valerius Terminus of the Interpretation of Nature: With the Annotations of Hermes Stella. In Vol. 3 *of The Works of Francis Bacon*, eds. James Spedding, Robert Leslie Ellis, Douglas Denon Heath. London: Longman & Co.

J. Hattiangadi, *Francis Bacon's Skeptical Recipes for New Knowledge*,
https://doi.org/10.1007/978-3-031-52585-8

———. (1620) 1858. Novum Organum. In Vol. 4 *of The Works of Francis Bacon*, 14 vols, ed. J. Spedding, R.L. Ellis, and D.D. Heath. London: Longman & Co.

———. (1609) 1884. The Wisdom of the Ancients. In *Bacon's Essays and the Wisdom of the Ancients*, ed. Basil Montagu. Boston: Little, Brown, and Company.

———. (1601) 1951. The Masculine Birth of Time. In *'Temporis Partus Masculus: An Untranslated Writing of Francis Bacon'* tr. Benjamin Farrington. *Centaurus* 1 (3): 193–205.

———. (1597) 1985. The Essays. In *The Essayes or Counsels, Civill and Morall*, ed. Michael Kiernan. Vol. XV of *The Oxford Francis Bacon*. Oxford: Clarendon Press.

———. (1605) 2000a. The Advancement of Learning. In *The Advancement of Learning*, ed. Michael Kiernan. Vol. IV of *The Oxford Francis Bacon*. Oxford: Clarendon Press.

———. (2026) 2000b. A New Abecedarium of Nature. In *The Instauratio Magna: Last Writings*, ed. Graham Rees. Vol. XIII of *The Oxford Francis Bacon*. Oxford: Clarendon Press.

Bernard, Claude. (1864) 1957. *An Introduction to the Study of Experimental Medicine*. Tr. H.C. Greene. New York: Dover,

Bett, Richard. 2000. *Pyrrho, his Antecedents, and his Legacy*. Oxford: Oxford University Press.

Boas, Marie. 1952. The Establishment of the Mechanical Philosophy. *Osiris*. 10: 412-541.

Boyle, Robert. 2002. *Correspondence of Robert Boyle*. 6 vols. Ed. Michael Hunter, Antonio Clericuzio, Lawrence M. Principe. London: Routledge.

Broad, Charles D. 1926. *The Philosophy of Francis Bacon*. Cambridge: Cambridge University Press.

Carson, Emily. 2006. Locke and Kant on Mathematical Knowledge. In *Intuition and the Axiomatic Method*, ed. Emily Carson and Renate Huber. Dordrecht: Springer.

Charron, Pierre. 1601. *De la Sagesse*. 3 vols. Bordeaux: Simon Millanges. English Edition: Charron, Pierre. 1697. *Of Wisdom* (trans. Stanhope, George). 3 vols. London: J. Tonson et al.

Cicero. 2006/45BCE. *On Academic Scepticism*. Tr. Charles Brittain. Indianapolis: Hackett.

Cohen, L. Jonathan. 1980. Some Historical Remarks on the Baconian Conception of Probability. *Journal of the History of Ideas*. 41 (2): 219-31.

Crawford, Peter. 2014. *The War of the Three Gods. Romans, Persians and the Rise of Islam*. New York, NY: Sky Horse Publishing.

Descartes, René. 1983. *Oeuvres de Descartes*, 11 vols. Ed. Charles Adam, Paul Tannery. Paris: Librarie Philosophique J. Vrin.

———. 1984. *The Philosophical Writings of Descartes*, vol. 2. Ed. John Cottingham, Robert Stoothoff, Dugald Murdoch. Cambridge: Cambridge University Press.

———. 1985. *The Philosophical Writings of Descartes*, vol. 1. Ed. John Cottingham, Robert Stoothoff, Dugald Murdoch. Cambridge: Cambridge University Press.

———. 1991. *The Philosophical Writings of Descartes*, vol. 3. Ed. John Cottingham, Robert Stoothoff, Dugald Murdoch, Anthony Kenny. Cambridge: Cambridge University Press.

———. (1664) 1998. *The World, or the Treatise on Light*. In *The World and Other Writings*, tr., ed. Stephen Gaukroger. Cambridge: Cambridge University Press.

Duhem, Pierre. (1906) 1954. *The Aim and Structure of Physical Theory*. Originally, *La Théorie Physique, son objet et sa structure*. Tr. Philip P. Wiener. Princeton, NJ: Princeton University Press.

Eddington, Arthur. 1949. *The Philosophy of Physical Science*. Cambridge: Cambridge University Press.

Einstein, Albert. 1905a. Zur Elektrodynamik bewegter Körper. *Annalen der Physik* 17 (10): 891–921.

———. 1905b. Ist die Trägheit eines Körpers von seinem Energieinhalt abhängig? *Annalen der Physik* 18 (13): 639–41.

Einstein, Albert, and Leopold Infeld. 1938. *The Evolution of Physics*. London: Cambridge University Press.

Feyerabend, Paul. 1975. *Against Method*. London: Verso.

Frede, Michael. 1987. *Essays in Ancient Philosophy*. Minneapolis: University of Minnesota Press.

Frege, Gottlob. (1884) 1953. *The Foundations of Arithmetic: A Logico-mathematical Enquiry into the Concept of Number*. Tr. J.L. Austin. Oxford: Blackwell.

Fresnel, Augustin. (1818) 1900. Memoir on the Diffraction of Light. In *The Wave Theory of Light: Memoirs by Huygens, Young and Fresnel*, ed. Henry Crew, 79–144. New York, N.Y.: American Book Company.

Galen. 1985/2nd c. *On Sects for Beginners*. Translated by Michael Frede. In *Three Treatises On the Nature of Science*, ed. Richard Walzer, Michael Frede. Indianapolis, Indiana: Hackett.

Galilei, Galileo. (1632) 1953. *Dialogue Concerning the Two Chief World Systems*. Tr. Stillman Drake. Berkeley: University of California Press.

———. (1623) 1957. The Assayer. In *Discoveries and Opinions of Galileo*, tr. & ed. Stillman Drake. New York: Doubleday.

———. (1638) 1974. *Two New Sciences*. Tr. & ed. Stillman Drake. Madison, WI: University of Wisconsin Press.

Gaukroger, Stephen. 2001. *Francis Bacon and the Transformation of Early-Modern Philosophy*. Cambridge: Cambridge University Press.

Hacking, Ian. 1983. *Representing and Intervening*. Cambridge: Cambridge University Press.

Hanson, Norwood Russell. (1958) 1965. *Patterns of Discovery: An Inquiry into the Conceptual Foundations of Science*. Cambridge: Cambridge University Press.

Harrison, Peter. 2023. Normativity and the Critical Functions of Genealogy: The Case of Modern Science. *Modern Theology*. 39 (4) 682-707. https://doi.org/10.1111/moth.12857.

Harvey, William. (1628) 1909. *On the Motion of the Heart and Blood in Animals*. Tr. Robert Willis. Vol. 38, Part 3 in *The Harvard Classics*, ed. Charles W. Eliot. New York, N.Y.: P.F. Collier & Son.

Heather, Peter. 2018. *Rome Resurgent: War and Empire n the Age of Justinian*. New York, NY: Oxford University Press.

Henry, John. 2002. *Knowledge Is Power. Francis Bacon and the Method of Science*. Cambridge: Icon Books.

Hesse, Mary B. 1965. Francis Bacon. In *A Critical History of Western Philosophy*, ed. Daniel John O'Connor. New York: Free Press.

Hume, David. (1748) 1993. *An Enquiry Concerning Human Understanding*. Ed. Eric Steinberg. Indianapolis, Indiana: Hackett.

Huygens, Christiaan. (1973) 1986. Christiaan Huygens' the Pendulum Clock, Or, Geometrical Demonstrations Concerning the Motion of Pendula as Applied to Clocks. In *Horologium oscillatorium dive de motu pendulorum ad horologia aptato demonstrationes geometricae*, tr. R. Blackwell. Ames, Iowa: Iowa State University Press.

Jalobeanu, Dana. 2015. *The Art of Experimental Natural History: Francis Bacon in Context*. Bucharest: Zeta Books.

———. 2019. Francis Bacon on Sophists, Poets and Other Forms of Self-Deceit (Or, What Can an Experimental Philosopher Learn from a Theory Informed History of Philosophy?) *Experiment, Speculation and Religion in Early Modern Philosophy*, ed. A. Vanzo, P.R. Anstey, New York, NY: Routledge.

Jardine, Lisa. 1974. *Francis Bacon: Discovery and the Art of Discourse*. Cambridge: Cambridge University Press.

Jardine, Lisa, and Alan Stewart. 1998. *Hostage to Fortune: The Troubled Life of Francis Bacon*. New York: Hill & Wang.

Joseph, H.W.B. 1916. *An Introduction to Logic*. 2nd ed. Oxford: Clarendon Press.

Kant, Immanuel. (1787) 1998. *Critique of Pure Reason*. Tr. Paul Guyer, Allen Wood. Cambridge: Cambridge University Press.

Keller, Evelyn Fox. 1985. *Reflections on Gender and Science*. New Haven, CT: Yale University Press.

Klein, Jürgen, and Guido Giglioni. (2020). 'Francis Bacon', *The Stanford Encyclopedia of Philosophy* (Fall 2020 Edition), Edward N. Zalta (ed.), https://plato.stanford.edu/archives/fall2020/entries/francis-bacon/.

Koyré, Alexandre. 1943. Galileo and Plato. *Journal of the History of Ideas* 4 (4): 400–428.

———. 1957. *From the Closed World to the Infinite Universe*. Baltimore, MD: Johns Hopkins Press.

———. (1938)1978. *Galileo Studies*, tr. John Mepham. *Études Galiléenne*, (Atlantic Highlands, NJ: Humanities Press.

Kuhn, Thomas. 1962. *The Structure of Scientific Revolutions*. Chicago: University of Chicago Press.

———. 1963. The Function of Dogma in Scientific Research. In *Scientific Change: Historical Studies in the Intellectual, Social and Technical Conditions for Scientific Discovery and Technical Invention, from Antiquity to the Present*, ed. A.C. Crombie, 347–69. London: Heinemann.

Lai, Tyrone. 1989. How We Make Discoveries. *Synthese* 79 (3): 361–92.

Lalande, André. 1911. Sur quelques textes de Bacon et de Descartes. *Revue de Métaphysique et de Morale* 19 (3): 296–311.

Laudan, Laurens. 1966. The Clock Metaphor and Probabilism: The Impact of Descartes on English Methodological Thought, 1660–1665. *Annals of Science* 22: 73–104.

Long, A.A., and D.N. Sedley. 1987. *The Hellenistic Philosophers, Vol. I: Translations of the Texts*. Cambridge: Cambridge University Press.

Mallet, David. 1768. *The Life of Francis Bacon, Lord Chancellor of England*. London: Printed for A. Millar.

Matthews, Steven. 2008. *Theology and Science in the Thought of Francis Bacon*. London: Routledge.

de Montaigne, Michel. (1580) 1993. *The Complete Essays*. Translated by M.A. Screech. London: Penguin Classics.

Newton, Isaac. 1672. New Theory of Light and Colors, Sent by the Author to the Publisher from Cambridge February 6 1671/72. *Philosophical Transactions* 6 (80): 3075–87.

———. (1726) 1848. *Newton's Principia: The Mathematical Principles of Natural Philosophy*. Tr. Andrew Motte. New York: Daniel Adee.

———. 1959. *Correspondence of Isaac Newton, Volume 1 1661–1675*. Ed., tr. Herbert W. Turnbull. Cambridge: Cambridge University Press.

———. 1975. *The Correspondence of Isaac Newton, Volume 5 1709–13*. Ed,tr. A. Ruper Hall, Laura Tilling. Cambridge: Cambridge University Press.

———. 2012. *The Optical Papers of Isaac Newton*. 3 vols. Translated and edited by Alan E. Shapiro. Cambridge: Cambridge University Press. Pappus. 2011/4th century. Book 7 of the *Collection*. Edited and Translated by Alexander Jones. New York: Springer.

Pérez-Ramos, Antonio. 1988. *Francis Bacon's Idea of Science and the Maker's Knowledge Tradition*. Oxford: Clarendon Press.

Plato. 1892. *The Dialogues of Plato, Translated into English with Analysis and Introductions*. 5 vols. Translated by Benjamin Jowett. Oxford: Oxford University Press.

———. 1961. *The Collected Dialogues of Plato, Including the Letters*. Ed. Edith Hamilton, Huntington Cairns. New York: Pantheon Books.

———. 1997. *The Complete Works*. Edited by John Cooper. Indianapolis: Hackett.

Poincaré, Henri. 1901. *Électricité et optique. La lumière et les théories électrodynamiques; leçons professées à la Sorbonne en 1888, 1890 et 1899*. Paris: G. Carré et C. Naud.

———. 1905. *Science and Hypothesis*. Translated by W.J. Greenstreet. London: Walter Scott.

———. 1908. *Science and Method*. Translated by Francis Maitland. London: Thomas Nelson & Sons.

Popkin, Richard. 1960. *The History of Scepticism from Erasmus to Descartes*. New York: Humanities Press.

Popper, Karl. 1983. *Realism and the Aim of Science*. Ed. W.W. Bartley, III. London: Hutchinson.

———. (1959) 2002a. *The Logic of Scientific Discovery*. London: Routledge.

———. (1963) 2002b. *Conjectures and Refutations: The Growth of Scientific Knowledge*. London: Routledge.

Post, John. 1970. The Possible Liar. *Noûs*, 4 (4) 495-409.

Ptolemy. 1984/2nd c. *Ptolemy's Almagest*. Tr. G.J. Toomer. London: Duckworth.

Quine, Willard Van Orman. 1951. Two Dogmas of Empiricism. *Philosophical Review* 60: 20–43.

Rorty, Richard. 1979. *Philosophy and the Mirror of Nature*. Princeton, NJ: Princeton University Press.

Roth, Leon. 1937. *Descartes' Discourse on Method*. Oxford: Clarendon Press.

Russell, Bertrand. 1912. *The Problems of Philosophy*. London: Williams and Norgate.

Russell, Bertrand, and Whitehead, Alfred North. 1910–13. *Principia Mathematica*. 3 vols. Cambridge: Cambridge University Press.

Sabra, A.I. 1967. *Theories of Light from Descartes to Newton*. London: Oldbourne.

Schuster, John. 1980. Descartes' *Mathesis Universalis*, 1619–28. In *Descartes: Philosophy, Mathematics and Physics*, ed. Stephen Gaukroger. Sussex: Harvester Press.

Sextus Empiricus. 1990/2nd c. *Outlines of Pyrrhonism*. Tr. R.G. Bury. Amherst, NY: Prometheus Press.

———. 1995/2nd c. *The Skeptic Way. Sextus Empiricus's Outlines of Pyrrhonism*. Tr. Benson Mates. New York, NY: Oxford University Press.

———. 2000a/2nd c. *Outlines of Scepticism*. Translated by Julia Annas and Jonathan Barnes. Cambridge: Cambridge University Press.

———. 2000b/2nd c. *Against the Ethicists*. Translated by Richard Bett. Oxford: Clarendon Press.

Shapiro, Alan E. 1990. *Before Newton: The Life and Times of Isaac Barrow*. Cambridge: Cambridge University Press.

Shapiro, Alan E. 2021. Newton Did Change His Views from Certainty to Probability. *Philosophy of Science* 88 (1): 169–180.

Soranus. 1956/2nd c. C.E. *Gynaecology*. Translated by Owsei Temkin. Baltimore, MD: Johns Hopkins University Press.

Spedding, James, ed. 1861–74. *The Letters and the Life of Francis Bacon*, 7 vols. London: Longman, Green, & Co.

———, ed. 1878. *An Account of the Life and Times of Francis Bacon, Extracted from the Edition of His Occasional Works*. London: Trubner.

Sprague, Rosamund Kent. 1972. *The Older Sophists*. Columbia: University of South Carolina.

Sprat, Thomas. 1667. *The History of the Royal Society of London, for the Improving of Natural Knowledge*. London: T.R.

Tarrant, Harold. 1985. *Scepticism or Platonism?* Cambridge: Cambridge University Press.

Urbach, Peter. 1982. Francis Bacon as a Precursor to Popper. *British Journal for the Philosophy of Science* 33 (2): 113–32.

Vickers, B. 1968. *Francis Bacon and Renaissance Prose*. Cambridge: Cambridge University Press.

Vlastos, Gregory. 1994. The Socratic Elenchus: Method Is All. In *Socratic Studies*, ed. Myles Burnyeat, 1–38. Cambridge: Cambridge University Press.

Vogt, Katja Maria. 2012. *Belief and Truth: A Skeptic Reading of Plato*. New York: Oxford University Press.

Walsh, Kirsten. 2021. Newton's Scaffolding: The Instrumental Roles of His Optical Hypotheses. In *Experiment, Speculation and Religion in Early Modern Philosophy*, ed. Alberto Vanzo, Peter Anstey. New York, NY: Routledge.

Woodward, James. 2003. *Making Things Happen*. Oxford: Oxford University Press.

Xenophon. 1897. *Memorabilia*. In Vol. 3 of *The Works of Xenophon*, ed. H.G. Dakyns. London: Macmillan & Co.

Zagorin, Perez. 1998. *Francis Bacon* Princeton, NJ: Princeton University Press.

Index[1]

A
Abbasid Caliphate, 159
Abecedarium of Nature, 38
Abstraction, v, 42, 134, 135, 168, 169, 187, 188
Abstraction as inductions, 10
Academic skepticism, 66, 69, 90, 91, 99
Active fallibilism, 32, 53–70, 160
Adam's Fall, *see* Fall of Adam
Advancement of Proficience and Learning, 100, 103, 121, 166, 168, 173
Aenesidemus, modes of, 86, 149
Affirmative general principles, knowledge of, 52
Agrippa, modes of, 86
Alchemy, vii, 110–112, 131, 152, 164, 183n2
  alchemical, vii, 105, 106, 110–112, 129, 131, 152, 164, 183n2, 190
Alexander, H. G., 79, 257
Anstey, Peter, 153n1
Anticipations of nature, xvii, 133, 150
Aporetic ending of dialogues, 55
Aporia, 39–41
Appearances, 20, 32, 68, 75, 78, 80, 85–88, 103, 137–140, 143, 147–152, 162, 165, 174, 177–179, 186, 192, 198, 205, 233, 249–251, 256, 270, 271
  living by, 32, 165
Apraxia, of Academic skepticism, 65, 66, 95
Arcesilaus, 65–68, 75–77, 87, 89

[1] Note: Page numbers followed by 'n' refer to notes.

J. Hattiangadi, *Francis Bacon's Skeptical Recipes for New Knowledge*,
https://doi.org/10.1007/978-3-031-52585-8

Aristotle, v–vii, 2, 13, 25, 42, 43, 52, 63, 64, 67, 86, 104, 105, 107, 110, 121, 127–129, 134, 135, 140, 147, 148, 162–164, 169, 171, 187–190, 193, 195, 200, 202, 203, 209, 215, 224
  induction (*see* Abstraction)
Astronomy, ix–xi, 76, 128, 129, 152, 185, 198, 208, 209, 271n2
Ataraxia, 74, 76, 77, 79, 81–83, 90, 119, 137, 177
Aubrey, John, 34–37, 39
Augustine, Bishop of Hippo, 83, 95–97, 142, 160, 217
Axioms, 6, 15, 17, 55, 118, 130, 134, 150, 161, 163, 175, 182

B

Bacon, Francis, v, 2, 31, 53, 85, 117, 133, 156, 186, 215, 251
  life of, 33, 35, 108
  works of, 69, 110, 119, 131, 146, 179, 182n2
Beeckman, Isaac, 219
Belief, privileged in faith over knowledge, 96–98
Bernard, Claude, 82
Bett, Richard, 73, 92n1
Boyle, Robert, xi, xii, xvi, 1, 4, 9, 106, 153n1, 160, 187, 210, 230, 231, 234, 249, 260, 261, 264
Bribes, bribery, 33–37
Buddhism, 79

C

Carneades, 75, 77, 88, 89
Carson, Emily, 29n1
Cartesian ethics, in the Discourse, 218
Cartesian gap in knowledge of things, 227
Causes and constant conjunctions, 8, 11, 14
Certainty derived from the *cogito*, 214
Certainty of knowledge in mathematics, Galileo, 219
Charity as the aim of science, 106, 182
Charron, Pierre, 213
Chemical cosmology, 127
Christian philosophy, *see* Subjugated philosophy
Church of England, 92, 100, 102, 106, 156, 166, 167
Church of Rome, *see* Roman
Cicero, 67–69, 77, 89, 95, 96, 99
Circularity in God and reason, Cartesian, 225
Clear and distinct ideas, 226, 247
Closure, epistemic, xx
Cogito, 214, 215, 221, 224, 226–228, 231, 258
Cohen, L. Jonathan, 182n1
Colors
  spectrum of, 245, 246, 251, 256, 262, 266
  Newtonian analysis of, 231–232
Combinatorial theory of Baconian induction, 126
Comets, debate between Sarsi and Galileo, 205

Composition and decomposition of motions, 201, 237
Conjectures, vi, viii, 51, 135, 138–140, 145, 147, 174, 178, 221, 268
Contemplative, combative persona of Socrates, 42
Context of discovery and justification, 4
Conventionalism, 15, 16, 19, 20
Copernicus, Nicolaus, 129
Craft, of skepticism, 61
Craftspeople, 43, 45, 113n4
Crawford, Peter, 158
Criterion
  of knowledge, 91, 97, 98, 165, 169, 216, 217
  of truth, 97, 216, 217
Criterion of existence, 63
  Sophist, 63

D

Dark room colors in monochrome light, 264
Decomposition of motions, *see* Composition and decomposition of motions
Demurge, in Plato's *Timaeus*, 104, 128, 198, 234
Descartes, René, 20, 195, 202, 204, 207, 210, 210n2, 213–221, 223–232, 234–238, 240–245, 247n1, 258, 260
  reductio ad absurdum of skepticism, 58
Dialectic, 59, 62, 101, 190, 198
Dialogues, Galileo, 204
Digamber Jains, *see* Jain religion
Discovery, of hidden reality, xiv
  See also Context of discovery and justification
Discrepancies experienced, 233
Discrepancies in the modes of Aenesidemus, 86
Distortion of reality, 144, 179
Divided line, Plato's, 196–198
Divine mathematical knowledge, *see* Certainty of knowledge in mathematics, Galileo
Duhem, Pierre, xiv–xvii, 18, 20–28, 195, 200, 257

E

Eddington, Arthur, 13, 14
Eidola in Plato, and Baconian idols, 108, 109
Einstein, Albert, 19, 27, 178, 261
Elements of cosmology, 129
Elenchus, Socratic, Bacon's, vii, 32, 42, 48–56, 58, 60, 62, 67, 92, 124, 125, 155–182
Elenctic, 54, 55
Eliminative induction, 122, 123, 125, 182n1
Emblematic ideologies, 157, 158
Empiricism, xi, 29n1, 81, 117, 204
Epistemology, 29n1, 67, 84, 130, 162, 186, 193, 194, 204, 214, 217
Epochē, 76, 77, 82, 83, 90, 213
Equipollence, 76
Errors, xiii, 3, 32, 45, 64, 68, 84, 96, 136, 138, 143, 145, 146, 151, 161, 163, 165, 169, 173, 174, 177–179, 181, 182, 190–194, 205, 206, 219–222, 228, 231, 249, 252, 256, 261, 267, 271

Essences, vi, vii, 13, 110, 141, 148, 162, 163, 168, 189, 200, 224, 225, 234
Etiology, medical, 80, 81, 139
Evidence, vii, viii, xii, xxvn2, 5, 7, 25, 36, 64, 73, 83, 88, 113, 118, 126, 133–153, 159, 160, 174, 178, 179, 181, 205, 218
Evil daemon hypothesis, 227
Experiment, vii, ix, x, 17, 18, 21–25, 27, 103, 106, 111, 112, 113n4, 117–121, 127, 130, 134, 138, 147–149, 151, 152, 153n1, 164, 167, 174, 175, 179, 182, 183n2, 185, 186, 190, 192, 201, 204, 213, 214, 220, 224, 227, 236, 241, 250, 252, 254–259, 261–269
Experimental philosophy, ix–xii, 3, 9, 11, 26, 29, 152, 186, 187, 230–233, 249–271
Explanation, 2, 38, 104, 123, 147, 182n2, 207, 229, 231, 233, 242, 258, 259, 263

F

Faith *vs.* knowledge, 91, 101, 209
Fallibilism/fallibilist, 31, 32, 53–70, 160
Fall of Adam, 103, 167, 171
Falsificationism, 135–136, 146
Fermat, Pierre, 231, 241, 260
Feyerabend, Paul, xix, xx
Forms, *see* Ideas, Platonic
Foundational skepticism *vs.* Pyrrhonian doubt, 102
Four causes of scholasticism, 105
Frede, Michael, 92n3
Frege, Gottlob, 5
Fresnel, Augustin, 26, 270, 271

G

Galen, 80
Galilei, Galileo, ix, x, 20, 21, 28, 51, 128, 185, 186, 191, 194–196, 199–205, 207–210, 213, 214, 219–222, 225, 230, 232n1, 237
  on divine and human knowledge, 186
  and the experimental method, 185, 210
  mathematical universe, 200
Gap of Cartesian principles from experience, 231
Gathering errors in natural histories, 117, 174
Gaukroger, Stephen, 182n1
Geometry of motion, 201, 203, 219
  non-Euclidean, 15–20
  See also Trajectories
Giglioni, Guido, 101
Gorgias, 42, 49, 50, 54–56, 58, 60
Gravity, x, 10, 14, 26–28, 257, 261, 265
Guarantee of our knowledge, God's, 225

H

Hacking, Ian, 182n2, 194
Hanson, Norwood Russell, 136, 144
Harrison, Peter, 70n1, 113n1
Harvey, William, xxvn2, 225
Heather, Peter, 156

Henry, John, 113n4
Hesse, Mary, 127, 131
Holism, 24
Hooke, Robert, 153n1, 234, 260, 261, 264, 268–270
  colored jars, 234, 268, 270
Hume, David, x, xi, xv, 6–17, 25, 29n1, 213, 223
Humean critique of induction, xv
Humility, Socratic, 78
Huygens, Christiaan, 231, 232n1, 242, 243, 260, 267, 270, 271
Hypotheses, method of, xii, xiv, xvi, 29, 42, 50, 51, 54, 61, 62, 117, 135, 145, 147, 213, 214, 234, 243, 249, 251
Hypothetical models, 199, 201, 210, 229, 231, 233, 234, 243, 244, 247, 251
  Cartesian, 231, 260
Hypothetico-deductive method, 4, 230, 260

I

Ideas, v–vii, xviii–xx, 4, 10–14, 22, 32, 34, 35, 39, 42, 57, 59, 60, 63, 64, 68, 70n2, 79, 88, 97, 100, 101, 105, 107–109, 112, 113, 122, 131, 136–140, 145, 148, 149, 151, 159, 161–163, 170–174, 176, 177, 179, 180, 182, 182n2, 186, 188, 191, 198, 201, 218, 224–227, 230, 234, 237, 247
  Platonic, 64
Identification with Socratic method, 31, 32
Idols of the mind, 108, 109, 140, 143, 149, 151
Ignorance, Socratic, 60, 62, 78
Imprisoned faith in Anglicanism, 100
Inclined plane, 185
Incommensurability, 22, 23
Index of refraction, 240, 246
Induction
  critique, xv
  impossibility of, 27, 122
  prospective, xvi, 4
  retrospective, xvi
  in Socrates' thought, vi, 42, 48
Inductive inference, xvi, 3, 5, 6, 25, 121
Infeld, Leopold, 27, 178
Intellectual midwife, 171
Interpretation of nature, vii, viii, 122, 134, 137, 145, 186, 263, 265
Islamic philosophy, *see* Subjugated philosophy
Isosthenia, see Equipollence

J

Jain religion, 79, 89
Jalobeanu, Dana, 113n3, 177
Jardine, Lisa, 38, 52n1, 182n1
Joseph, H. W. B., 122, 124, 175

K

Kant, Immanuel, 11–13, 15–17, 21, 223
Keller, Evelyn Fox, 39
Khalifates, 159

Kinematics, 199, 200, 203, 204, 237
Klein, Jürgen, 101
Knowledge
 affirmative, 48–52, 58
 negative, 78, 171
 skeptical knowledge, xiii, 31, 78
Koyré, Alexandre, 194–196, 199, 200
Kuhn, Thomas, xviii, xix, 178, 193

L

Lai, Tyrone, 121, 178
Lalande, André, 219
Law of fall, Galileo's, 200
Law of refraction, Cartesian, *see* Snell, Willebrord
Learning from error, 267
Leibniz, Gottfried Wilhelm von, 223, 257, 259
Light as a particle, *see* Hypothetical models
Light as a pressure, *see* Hypothetical models
Light as instantaneous motion, *see* Hypothetical models
Living by appearance, its practicality, 165
Local motions, 199, 200, 207
Locke, John, 29n1, 135, 141, 173, 191, 223
Logic
 deductive logic, 4, 5, 134
 inductive, v, vi, xvi, 4, 5, 11
Long, A. A., 68

M

Maeutics, 39–43
Mallet, David, 52n1
Masculine birth of time, 38, 39, 42, 109, 171, 172, 191
Mathematics, 14, 16–19, 21, 29n1, 58, 61, 62, 197–201, 204, 208, 209, 218, 219, 221, 222, 228
 See also Method, of hypothesis in mathematics
Matters of fact, 7, 16
 operation of thought, 7
Matthews, Stephen, 102
Meno, 43, 49–51, 54–62, 78, 135, 147, 169, 170, 180, 216
Metaphysics, 2, 20, 21, 106, 138, 145, 162, 193, 195, 200, 202, 207, 208, 224
Method
 of hypothesis in mathematics, 62
 retrospective and prospective
 of science, vii, xii, xiv, xx, 1, 6, 32, 48, 69, 112, 178, 219, 220, 233
 of science supersedes any theory, 261
Methodism in medicine, 81
Methodology, Cartesian, 220–228
Model, 85, 120, 131n1, 179–182, 193, 194, 231, 249, 259
 Cartesian (*see* Hypothetical models)
 of light, in the Cartesian optics, 242
 schematism of reality, Baconian, 85
Modes of Aenesidemus, 86, 149
Modes of Agrippa, 86
Monotonic inferences, 175, 182n1
Montaigne, Michel de, 98, 99, 108, 113n3, 165, 213, 218

N

Natural history, Baconian, ix, xx, 174, 178
Natures, vii–ix, xvii, xx, 2, 5, 7–12, 16, 24, 25, 32, 39, 40, 44, 47, 48, 52, 67, 68, 78, 80, 86, 88–90, 101–107, 109–113, 113n4, 118–122, 124–128, 130, 133–135, 137, 138, 140–143, 145–153, 153n1, 156, 161–164, 167, 168, 171–182, 183n2, 186, 189–192, 194, 200–203, 205, 215, 229, 230, 234, 243, 250, 256, 259, 263, 265, 266
Necessity as a mark of knowledge, 186–188, 201–202
Negative dogmatism, of Academics, 61, 74, 77, 90
Newton, Isaac, vi, ix–xii, xv, xvi, xx, xxiii, xxiv, 1–4, 9, 10, 12–14, 20, 25–29, 121, 137, 152, 153n1, 178, 187, 205, 230–233, 232n1, 240, 242, 249–271, 271n1, 271n2
  color wheel, 266, 267
  laws contradict Galileo's, 28
  laws contradict Kepler's, 28
  Opticks, 257
  Principia Mathematica, 205, 257–259
  reflecting telescope, 250, 256, 258
Non-Euclidean Geometries, 15–20
Non-hypothetical knowledge, *see* Unconditional knowledge
Novum Organum, 39, 107, 118, 127, 134, 140, 143, 145, 147

O

Operation of thought, *see* Matters of fact
Optics, ix, 3, 23, 137, 152, 225, 230–232, 234, 237, 243, 245, 249–252, 257, 259, 262, 267, 271n2
Organon, see Novum Organum
Orthodoxy, xix, 96, 97, 99, 113n1, 155–160

P

Pappus, 221
Parmenides, 59, 63, 64, 162, 196
Partial interpretation of reality, *see* Distortion of reality
Patents. letters of patent, 107
Peace of mind, *see Ataraxia*
Peacock's feathers, colors of, 264
Pederasty
  Bacon's, 35, 37, 38
  Greek, 37
Peirce, Charles, S., xii, 182n2
Pendulum motion, 201
Perception, 13, 14, 68, 83–85, 110, 136, 140, 142, 143, 145, 147, 148, 151, 171, 173, 189–191, 197, 253
Pérez-Ramos, Antonio, 113n4, 182n2
Personal knowledge for Socrates, 39, 43–45, 48, 50, 51, 56–62, 70, 78, 89, 169, 170, 175, 196, 198, 203, 216
Phaedo, 59, 63, 170

Philosophy, v, vii–xiv, xviii–xx, 2, 3, 32, 53, 73, 96, 117, 137, 155, 185, 233, 249–271
science, v, xii–xiv, xix, xx, 3, 4, 6, 13, 15, 19, 22, 182n2, 260
Physics, vi, ix, x, xiii, xviii, 14, 17–28, 79, 92n3, 105, 106, 185–210, 219, 222, 225, 237, 240
Physiologic, in the history of medicine
Plato, xiv, xxiii, xxiv, 31, 32, 38, 39, 42–46, 48–51, 53–70, 70n2, 70n3, 70n4, 74, 77–79, 85, 89, 103–105, 108, 109, 113n1, 113n2, 122, 124, 125, 128–130, 135, 136, 142, 143, 147, 151, 156, 160–163, 169–172, 182, 187, 188, 191, 196–200, 203, 204, 209, 216, 217, 234
Gorgias, 42, 50, 54, 55, 58
Meno, 43, 49–51, 54–62, 78, 135, 147, 169, 216
Phaedo, 59, 63, 170
Republic, 63, 64, 67, 109, 170, 196, 197, 199
Sophist, 63, 109
Theaetetus, 39, 42, 67, 68, 135, 136, 143, 151, 170, 171
on the unreality of motions and change, 128, 188, 198–200
Plea bargain, Bacon's, 37
Poincaré, Henri, xiv–xvi, 15–21, 23–27
Popkin, Richard, 91, 95, 97, 99, 164, 165
Popper, Karl, xvi–xviii, 28, 50, 135, 136, 138, 141–147, 178
Post, John, 70n3
Power over nature, 32, 102, 103, 106, 146, 153, 181, 182, 194, 250, 256, 266
Prism, 245, 246, 250–252, 254, 255, 262, 263, 265, 268, 269
Ptolemy, Claudius, 237
Puritans, 102
See also Church of England
Puzzle, viii, ix, xx, 120, 121, 138, 143, 145, 175, 178–181, 193, 194, 205, 249, 260, 261, 265, 266
Pyramid of science, 161, 261
Pyramid removed from top of pyramid, 104–106
Pyrrhonian skepticism, 73, 75, 80, 83, 91, 92n1, 151, 174

Q

Quests, 12, 59, 74
Quine, Willard Van Orman, 24, 176

R

Rainbow, as explained by Descartes, 244, 245
Rational, 3, 5, 6, 8–11, 14, 17, 117, 148, 155, 189, 215–217, 223, 225, 226, 228, 231, 233, 259
reconstruction, xvi, 4
Rationalism, refuting ancient skepticism, 214–216
Recipes for knowledge, 4
Rectilinear propagation of light, 270
Reductio ad absurdum and the *elenchus*, 58

Reduction to absurdity, *see* Thought experiments
Reflecting telescope, *see* Newton's, reflecting telescope
Refraction of light, at different angles to glass, 244
Refraction of light, through a second prism, 255
Refractive index, of materials, 262
Refrangibility of rays of light, 262
Refutation
  elenctic, 54, 55
  of skepticism, two, 138, 149, 151, 216
Refuting evidence, keeping a record of, vii, viii, 73, 137, 138, 174
Relativism, 49, 56, 85
Relativity, in the modes of Aenesidemus, 84, 86, 149
Republic, 63, 64, 67, 109, 156, 170, 196, 197, 199
Roman, 80, 102, 157–160
Roman Catholic Church, ix, 96
Rorty, Richard, 173
Royal Society of London, the, viii, 2, 53, 160, 230, 231
Russell, Bertrand, 5, 217

S

Sabra, A. I., 240
Sacred tradition of a Church, 92, 97, 165
Schematism, *see* Model
Schools of thought, 1, 65, 75, 105, 108, 150
Science, as knowledge, v, 17, 103, 203, 222
Science, method of, vii, xii, xiv, xx, 1, 6, 32, 48, 69, 112, 178, 219, 220, 233
Secular, 100, 101, 103, 105, 113n1, 155–157, 159, 160, 164, 167
Sedley, David Neil, 68, 92n2
Sextus Empiricus, 68, 69, 73–84, 86, 87, 89–91, 99, 136, 139, 149, 150, 215
Shadow, spot in the middle of, 26
Shapiro, Alan, 257, 258
Situational skepticism, 69, 76, 89
Skeptical knowledge, xiii, 31, 78
Skepticism
  academic, Pyrrhonian, 8, 65, 69, 73–92, 149–151, 165, 174, 176, 177, 216, 220
  as a method, xiii, 160
  selfrefuting, 62, 138
Snell, Willebrord, 219, 243
  sine law of reflection, 236–240
Socrates, vi, 31–62, 66–68, 70, 78, 79, 89, 140, 141, 160, 169–172, 189, 196, 198, 203, 216
Sola fide, 101–102
Sola scriptura, 92, 97, 98, 101, 160
Solving the whole optical puzzle, Newton, xx, 178, 193, 249, 261, 265, 266
Sophist, 43–47, 49, 55–57, 60, 63, 109, 124
Soranus, 80, 92n4, 92n5
Spedding, James, xxvn1, 52n1, 110
Spinoza, Benedict, 11, 180, 223
Sprague, Rosamund Kent, 49
Sprat, Thomas, ix
Stevin, Simon, 219
Stewart, Alan, 38, 52n1
Stoicism, 67, 124

Subjugated philosophy, *see* Submissive philosophy
Submissive philosophy, 159
Substances in Descartes' Discourse, three, 202, 207, 223, 224
Superposition of motions, *see* Composition and decomposition of motions
Sylva Sylvarum, 140
Symbolic interpretation of experiment, 22
Symptoms, 32, 78, 80, 81, 139

T

Tables of degrees of a nature, 119
Tables of presence and absence of a nature, 181
Tarrant, Harold, 69, 70n4
Temporis Partus Masculis, Bacon's, 38
Tests of knowledge, 92n5, 214
Theaetetus, 39, 41, 42, 67, 68, 70n2, 135, 136, 143, 151, 170, 171
Theories, v–vii, ix–xi, xvi, xvii, xx, xxvn2, 2, 3, 5, 6, 9, 11, 18–28, 31, 32, 44, 49, 50, 64, 69, 77, 81, 83, 85, 88–91, 111, 112, 120, 124, 125, 129–131, 134–136, 138, 141, 152, 168, 170, 186–188, 190, 193, 195, 198, 199, 201–203, 205, 210, 217, 220, 228, 230–232, 242–245, 251, 253, 257, 258, 260–262, 265, 268–270
  of colors as twists of light, 246
Theory-laden observations, theory-ladenness, 136, 151
Thought experiments, 201, 204, 213, 214, 220, 224, 227
Thought police, *see* Orthodoxy
Timaeus, 104, 128, 198, 234
Trajectories
  as geometrical, 199, 237
  and real, 199
Transverse waves of light, 270

U

Umayya, 159
Unconditional knowledge, 198–199
Urbach, Peter, 145, 146

V

Valerius Terminus, 100
Validity as a clear and distinct intuition, 213–218
Vickers, B., viii
Virtue, teachability of, 55
Vlastos, Gregory, 42, 48, 50, 54
Vogt, Katja Maria, 70n4

W

Wave theory of light, 26, 242, 269, 270
Whitehead, Alfred North, 5
Woodward, James, 182n2, 194
Works, and the aim of science, 28

X

Xenophon, 43, 46

Z

Zagorin, Perez, 109

Kolibri 360 GmbH
Ferdinand-Jühlke-Straße 7
99095 Erfurt, Deutschland
produktsicherheit@kolibri360.de